Functional Skills

Maths

Level 1

This CGP book covers everything you'll need for success in Level 1 Functional Skills Maths, whichever exam board you're studying.

Every topic is explained with clear, concise notes — and there's a huge range of practice questions and test-style tasks (with detailed answers) to help you make sure you're fully prepared for the final test.

Since 1995, CGP has helped millions of students do well in their tests and exams. Our books cover dozens of subjects for all ages — at the best prices you'll find anywhere!

Study & Test Practice

Contents

Section 1 — Number

Section 2 — Measure

Section 3 — Shape and Space

Section 4 — Handling Data

Test-Style Questions

Published by CGP

Editors:
Katie Braid, Christopher Lindle, Hayley Thompson.

Contributor:
George MacDonald.

Updated by Alex Fairer and Ben Train.

With thanks to Eva Cowlishaw, Rosie McCurrie and David Norden for the proofreading.

ISBN: 978 1 78294 632 8

Printed by Elanders Ltd, Newcastle upon Tyne.
Clipart from Corel®

Adding and Subtracting

You Need to Know When to Add or Subtract

1) The questions you get in the test will be based on real-life situations.
2) You won't always be told whether to add or subtract (take away).
3) You'll need to work out for yourself what calculation to do.

EXAMPLE 1:

Richard is in a restaurant. The restaurant menu is shown below.
Richard orders soup and roast beef. How much does his meal cost him?

Soup	£2
Roast beef	£9
Salmon	£10
Lemon tart	£3

Answer: you need to add together the price of the soup and the price of the roast beef. So the calculation you need to do is:

2 + 9 = **£11**

Sometimes you need to include units in your answer. Units tell you what type of number you've got. In this case the units are '£'.

EXAMPLE 2:

A rugby team wins 5 points from their first game of the season, 3 points from their second and 1 point from their third. They are then fined and lose 2 points.

How many points does the team have in total?

Answer: this calculation has two steps.

1) Add up the number of points the team wins:

5 + 3 + 1 = 9

2) Then take away the points the team loses:

9 – 2 = 7

The rugby team has **7** points in total.

Always Check Your Answer

1) Adding and subtracting are opposite calculations.
2) Once you've got your answer, you can check it using the opposite calculation.
3) You should get back to the number you started with.

EXAMPLE 1:

What is 300 – 102?

Answer: 300 – 102 = **198**

Use a calculator to work this out. You'll be able to take a calculator into the test and use it whenever you need to.

Check it using the opposite calculation: 198 + 102 = 300

EXAMPLE 2:

What is 36 + 54?

Answer: 36 + 54 = **90**

You only need to do one of these calculations to check your answer.

Check it using the opposite calculation: 90 – 54 = 36 OR 90 – 36 = 54

Practice Questions

1) Alice is in the supermarket. She buys chicken for £6, potatoes for £2 and carrots for £1.
How much does Alice spend in total?

....

2) Amir buys a drink from the newsagent for £1. He pays with a £5 note.
How much change should the newsagent give Amir?

....

3) Anna sells cosmetics door-to-door. Last week she had 10 customers.
This week, she loses 2 of her old customers but gets 3 new customers.
How many customers does she have this week?

....

4) Paulo has £500 in his bank account. He pays a bill for £346 by direct debit.
How much money will Paulo have in his account once the bill has been paid?
Show how you check your answer.

....

....

Multiplying and Dividing

You Need to Know When to Multiply or Divide

You'll get questions in the test where you need to multiply or divide. You're allowed to use a calculator but you'll need to be able to work out what calculation to do for yourself.

EXAMPLE 1:

Jim drives a total of 23 miles to and from work each day.
In January he works 21 days. How many miles does Jim drive in January?

Answer: Jim drives 23 miles each day for 21 days.
So you need to calculate 23 times 21.

23 × 21 = **483 miles** ← The units here are 'miles'.

EXAMPLE 2:

Louise is organising a trip for her students. The trip costs £280.
The cost of the trip will be split equally between 28 students.
How much will each student pay?

Answer: the £280 has to be divided between 28 students.
So you need to calculate 280 divided by 28.

280 ÷ 28 = **£10**

Always Check Your Answer

1) Multiplying and dividing are opposite calculations.
2) Once you've got your answer, you can check it using the opposite calculation.
3) You should get back to the number you started with.

EXAMPLE:

What is 54 × 6?

Answer: 54 × 6 = **324**

Check it using the opposite calculation: 324 ÷ 6 = 54 OR 324 ÷ 54 = 6

Practice Questions

1) Sarah is organising a trip. She has booked 5 minibuses. Each minibus seats 9 people.
How many people can go on Sarah's trip?

....

2) Jeremy sells his jam tarts in packs of 8. He has 64 jam tarts.
How many packs of jam tarts can he make altogether?

....

3) 11 × 33 = 363
Show how you could check this calculation is correct.

....

4) Stephanie sells chocolates in boxes of 9.

a) One day she makes 108 chocolates. How many boxes of chocolates can she make?

....

b) The next day she makes enough chocolates to fill 6 boxes exactly.
How many chocolates does she make on this day?

....

5) There are 28 rulers in a box. A school buys 33 boxes of rulers.
Will the school have enough new rulers for its 900 pupils?

....

6) James buys 3 boxes of jelly beans. There are 16 flavours of jelly bean.
Each box has 25 jelly beans of each flavour.

a) How many jelly beans are there in one box?

....

....

b) How many jelly beans has James bought in total?

....

Some Questions Need Answers that are Whole Numbers

1) Real-life division questions can be tricky.
You won't always end up with a whole number.

2) But sometimes, you'll need to give a whole number as your answer.

EXAMPLE 1:

Rick sells flowers in bunches of 6. He has 55 flowers.
How many bunches can he make altogether?

Calculation: $55 \div 6 = 9.17$

You can't have 9.17 bunches of flowers, so you need to give your answer as a whole number.

9.17 is between 9 and 10. There aren't enough flowers for 10 bunches.
So Rick will only be able to make **9 bunches**.

EXAMPLE 2:

Stella needs 42 chocolate buttons to decorate a cake.
The buttons are sold in packs of 15. How many packs should Stella buy?

Calculation: $42 \div 15 = 2.8$

Stella can't buy 2.8 packs of buttons, so you need to give your answer as a whole number.

2.8 is between 2 and 3. If Stella buys 2 packs of buttons, she won't have enough to decorate her cake. So Stella will need to buy **3 packs**.

Practice Questions

1) Daphne is making necklaces. Each necklace uses 7 large beads.
Daphne has 60 large beads. How many necklaces can she make altogether?

....

2) Simon buys 163 flowers. He wants to split the flowers into 7 equal bunches.
How many flowers will be in each bunch?

....

3) Nina makes 1500 g of strawberry jam. She needs to put it into jars.
Each jar holds 400 g. How many jars can Nina fill with jam?

....

Using a Calculator

Calculations with Several Steps

1) You'll sometimes need to do calculations that have several steps.
2) You could work out each step separately or you could type the whole thing into your calculator in one go.
3) BUT you need to be careful about how you type things into your calculator.

Some Calculators Use Brackets (...)

1) Some calculators use brackets to help them work out calculations with several steps.
2) The brackets tell the calculator to work out the bits inside the brackets before it does the rest of the calculation.
3) Without them, the calculator does the calculation in the wrong order — and you get the wrong answer.

EXAMPLE:

Nadia buys a drink and a bag of crisps for each of her 3 children. A drink costs 80p. A bag of crisps costs 43p. How much does Nadia spend in total?

1) You could work out the total cost of a drink and a bag of crisps for 1 child, then times this by 3.

   Total cost of a bag of crisps and a drink = 80 + 43 = 123p

   123 × 3 = 369p or **£3.69**

2) You could also use your calculator to do the whole calculation in one go.

   But if you type in '80 + 43 × 3', you might get the wrong answer (it depends on what kind of calculator you've got).

   You need to tell the calculator to work out the total cost of a drink and a bag of crisps for 1 child first, then times this by 3.

   So you need to put the 43 + 80 in brackets:

   (43 + 80) × 3 = 369p or **£3.69**

   Brackets always come in pairs.

Calculators Without Brackets

Not all calculators use brackets. You still need to be careful about how you type calculations into your calculator though.

EXAMPLE:

What is 6 divided by the answer to 5 × 2?

Work out what 5 × 2 is first, then divide 6 by this number: 5 × 2 = 10

6 ÷ 10 = **0.6**

If you just type '6 ÷ 5 × 2' into your calculator, you'll get the wrong answer:

6 ÷ 5 × 2 = 2

So if your calculator doesn't have brackets buttons, it's best to work out each step of the calculation separately.

If your calculator doesn't have bracket buttons and you're given a calculation that has brackets in, just work out the bits in brackets first.

EXAMPLE:

What is 18 ÷ (3 × 3)?

1) Work out the bit in brackets first: 3 × 3 = 9
2) Put this answer into the calculation instead of the brackets: 18 ÷ 9 = **2**

Practice Questions

1) What is 4 ÷ (2 × 5)?

....

2) What is (2 × 6) ÷ (3 × 2)?

....

3) What is (36 ÷ 6) ÷ (24 ÷ 12)?

....

4) Sam works a 12 hour shift on a Monday and a 6 hour shift on a Tuesday.
He does this every week for 4 weeks.
How many hours does Sam work in total over the 4 week period?

....

The Number Line and Scales

Negative Numbers are Less than Zero

1) A negative number is a number less than zero.
2) You write a negative number using a minus sign (-). For example, -1, -2, -3.
3) A number line is really useful for understanding negative numbers.

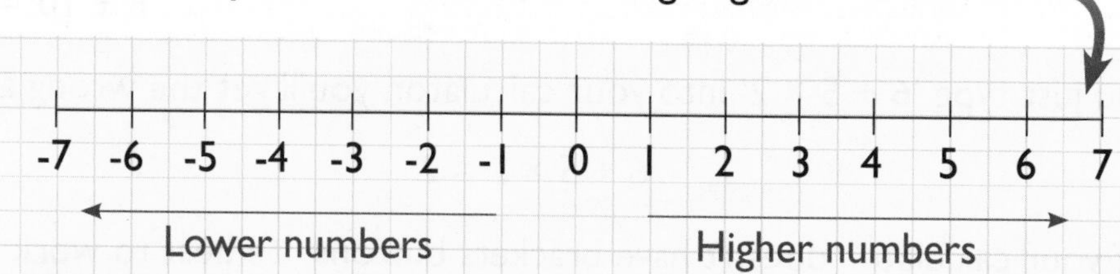

All negative numbers are to the left of zero.

All positive numbers are to the right of zero.

The further right you go, the higher the numbers get.
For example, -2 is higher than -7.

Use a Number Line to Work Out Differences

You can use a number line to work out the difference between two numbers.
For example, the difference between a negative number and a positive number.

EXAMPLE:

In London the temperature is -2 °C. In Paris it's 6 °C.

What is the difference in temperature between the two places?

1) Draw a number line that includes both the numbers in the question.
2) Count on from -2 to 6.

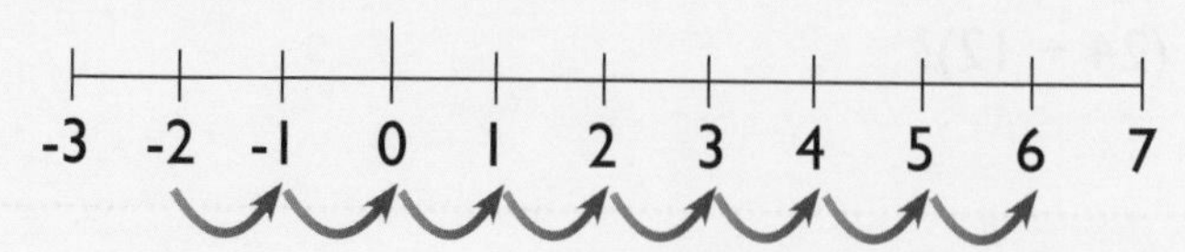

There are 8 steps, so the difference in temperature is **8 °C**.

Practice Questions

1) Which of these numbers is lower: -8 or -4?

....

....

2) In the morning it is -3 °C. At night it is -5 °C.
Is the temperature lower at night or in the morning?

....

....

3) The temperature outside Jan's house is -2 °C. Inside it is 21 °C.
What is the difference between these two temperatures?

....

....

A Scale is a Type of Number Line

1) You might be asked to read a scale.
For example, to read the temperature off a thermometer.

2) Scales are just number lines.
They don't always show every number though.

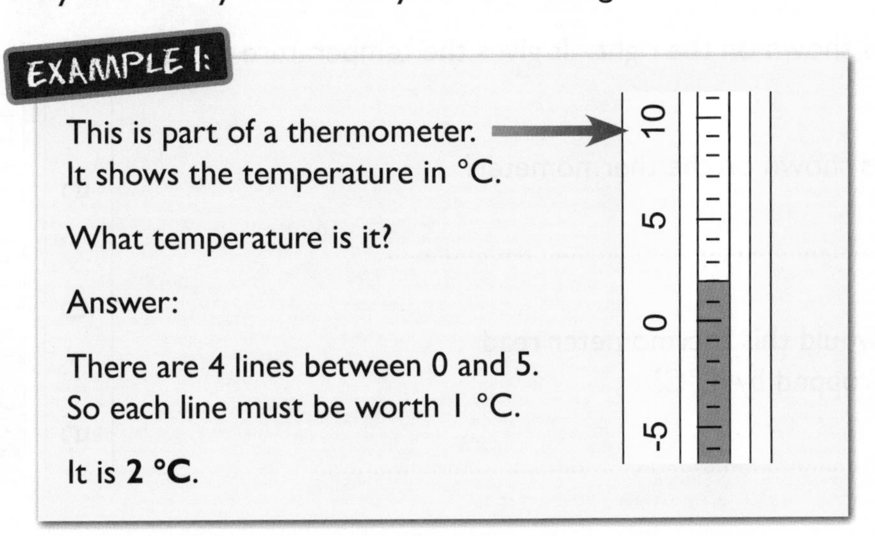

EXAMPLE 1:

This is part of a thermometer.
It shows the temperature in °C.

What temperature is it?

Answer:

There are 4 lines between 0 and 5.
So each line must be worth 1 °C.

It is **2 °C**.

EXAMPLE 2:

This is part of a thermometer.
It shows the temperature in °C.

What would the thermometer read if the temperature dropped by 6 °C?

Answer:

1) Work out what the thermometer reads now.

2) Count down 6 places.

The thermometer would read **-4 °C**.

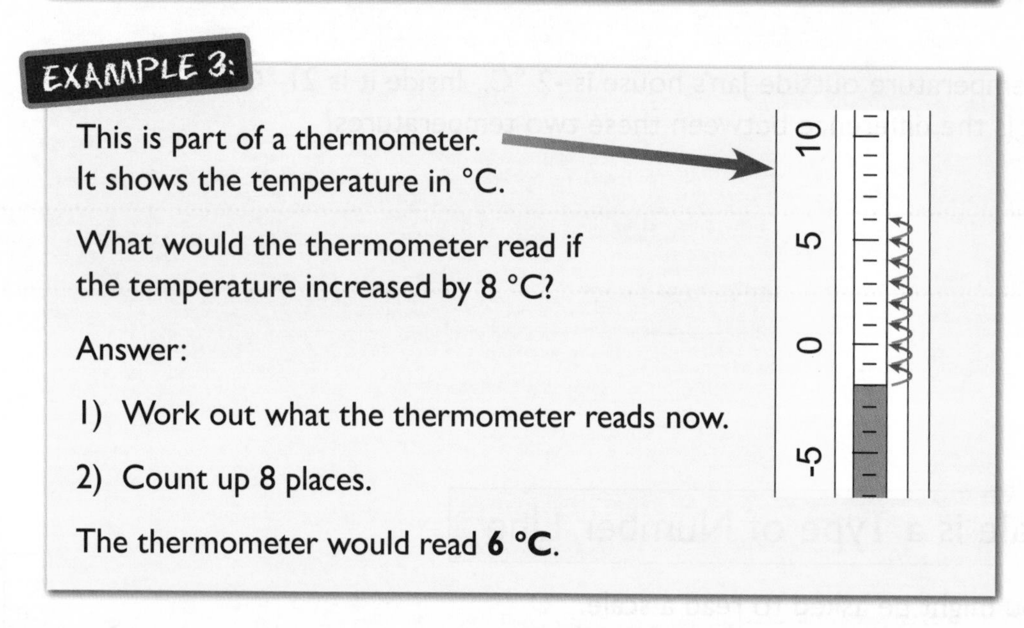

EXAMPLE 3:

This is part of a thermometer.
It shows the temperature in °C.

What would the thermometer read if the temperature increased by 8 °C?

Answer:

1) Work out what the thermometer reads now.

2) Count up 8 places.

The thermometer would read **6 °C**.

Practice Questions

Part of a thermometer is shown on the right. It gives the temperature in °C.

1) What temperature is shown on the thermometer?

....

2) What temperature would this thermometer read if the temperature dropped by 4 °C?

....

Fractions

Fractions Show Parts of Things

1) If something is divided up into equal parts, you can show it as a fraction.
2) There are two bits to every fraction:

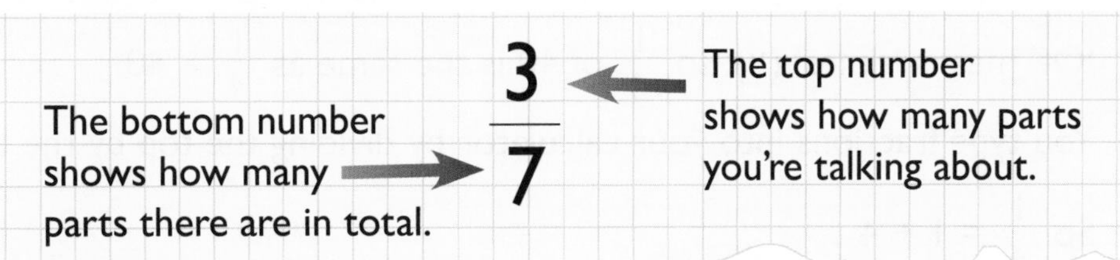

EXAMPLE:

Gemma has 7 squares of chocolate. She eats 4 squares.
What fraction did she eat?

She's eaten 4 out of the 7 squares, so it's $\frac{4}{7}$ (you say 'four sevenths').

Learn How to Write Fractions

Here's how to write some common fractions:

One half = $\frac{1}{2}$ One third = $\frac{1}{3}$

One quarter = $\frac{1}{4}$ Three quarters = $\frac{3}{4}$

You can also get mixed fractions. Mixed fractions are when you have whole numbers and fractions together. For example, $1\frac{1}{4}$ (one and a quarter).

Practice Questions

1) Jean's cat has 5 kittens. Jean gives 3 of the kittens away.
What fraction did Jean give away?

....

2) Write two and a half as a mixed fraction.

....

'Of' means 'times'

1) Sometimes, you might need to calculate a 'fraction of' something.
2) In these cases, 'of' means 'times' (multiply).

EXAMPLE 1:

What is $\frac{1}{4}$ of 40?

1) 'Of' means 'times' (×), so $\frac{1}{4}$ of 40 is the same as $\frac{1}{4} \times 40$.
2) You type fractions into your calculator by dividing the top by the bottom.

   So $\frac{1}{4} = 1 \div 4$

The overall calculation you need to do is: $1 \div 4 \times 40 =$ **10**

EXAMPLE 2:

A committee with 32 members takes a vote. Three quarters of the committee members vote 'yes'. How many members vote yes?

You need to calculate three quarters of 32.

1) 'Of' means 'times' (×), so $\frac{3}{4}$ of 32 is the same as $\frac{3}{4} \times 32$.
2) Type it into your calculator: $3 \div 4 \times 32 = 24$

   So **24** members vote yes.

Practice Questions

1) What is $\frac{1}{3}$ of 18?

   ....

2) What is $\frac{2}{5}$ of 50?

   ....

3) Keiran is a hotel receptionist. He works forty two hours a week. Keiran estimates that he spends a quarter of his time on the phone and a third of his time dealing with complaints.

   a) How many hours a week does Keiran spend on the phone?

   ....

   b) How many hours a week does Keiran spend dealing with complaints?

   ....

Discounts Involving Fractions

You need to be able to work out discounts involving fractions.

EXAMPLE 1:

A sofa usually costs £600. In the sale, it's half price.
What is the sale price of the sofa?

This is just like saying that the sofa costs 'half of £600' in the sale.

So you need to work out: $\frac{1}{2} \times 600$

Type it into your calculator: $1 \div 2 \times 600 =$ **£300**

EXAMPLE 2:

A bed usually costs £810. In the sale, there's a third off.
What is the sale price of the bed?

This time, you need to calculate £810 take away one third.

1) First you need to work out a third of £810. This is the same as $\frac{1}{3} \times 810$.

   Type it into your calculator: $1 \div 3 \times 810 =$ £270

2) Then you need to take this number away from £810: $810 - 270 =$ **£540**

Practice Questions

1) A tin of baked beans costs 68p. How much would the same tin cost if it was half price?

   ....

2) Cleo's car insurance would normally cost her £570. If she buys it online, she'll get a third off.
   How much will Cleo's car insurance cost with this discount?

   ....

   ....

3) Tim is booking a hotel room. The room he wants costs £120 per night.
   He finds a web site offering the same room for three quarters of the price.
   How much does the room cost on this web site?

   ....

   ....

Decimals

Not All Numbers Are Whole Numbers

1) Decimals are numbers with a decimal point (.) in them. For example, 0.5, 1.3.
2) They're used to show the numbers in between whole numbers.

EXAMPLES:

The number 2.1 is a bit bigger than the number 2.

The number 2.9 is a bit smaller than the number 3.

The number 2.5 is exactly halfway between the numbers 2 and 3.

3) You can show decimals on a number line.

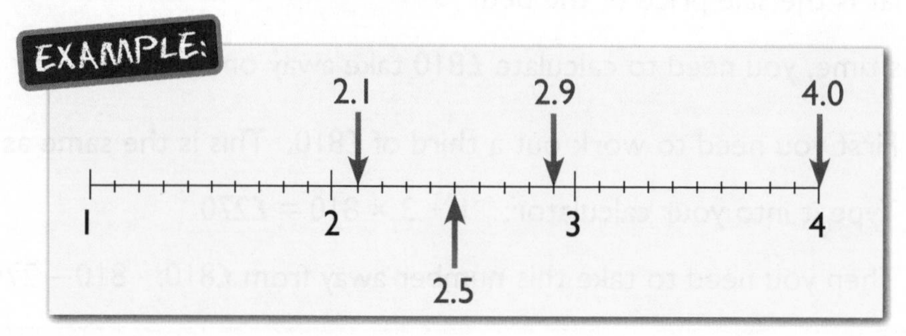

How to Put Decimals in Order

You might need to arrange a list of decimal numbers in order of size.

EXAMPLE:

Put these decimals in order of size: 1.7, 0.7, 0.07, 0.37. Start with the smallest.

1) Put the numbers into a column, lining up the decimal points.
2) Make all the numbers the same length by filling in extra zeros at the ends.
3) Look at the numbers before the decimal point.
   Arrange the numbers from smallest to largest.
4) If any of the numbers are the same, move onto the numbers after the decimal point. Arrange the numbers from smallest to largest.

Step 1:	→	Step 2:	→	Step 3:	→	Step 4:
1.7		1.70		0.70		0.07
0.7		0.70		0.07		0.37
0.07		0.07		0.37		0.70
0.37		0.37		1.70		1.70

The order is: **0.07, 0.37, 0.7, 1.7**

Adding and Subtracting Decimals

1) You can add and subtract decimals using a calculator.

2) It's exactly the same as with whole numbers — just remember to type the decimal point into the calculator.

EXAMPLE 1:

Sarah wants to know how much she spends on lunch each week.

Today she spent £2.75 on a sandwich and £1.32 on a cup of coffee. Yesterday she spent £1.60 on some soup.

How much has she spent so far this week in total?

Answer: add together everything Sarah has spent.

2.75 + 1.32 + 1.60 = **£5.67**

EXAMPLE 2:

Paul has been on a diet and lost some weight. He used to weigh 93.7 kg. He now weighs 88.4 kg. How much weight has he lost?

Answer: take away what Paul weighs now from what he used to weigh.

93.7 – 88.4 = **5.3 kg**

Practice Questions

1) Put these weights in order of size: 0.03 kg, 1.1 kg, 0.6 kg. Start with the smallest.

....

2) Jenna wants to know how much her baby daughter Lucy has grown.
Lucy was 48.75 cm long. She is now 50.6 cm long. How much has Lucy grown?

....

3) Matt goes to the garden centre. If he spends £20 or more he will get a free watering can.
Matt spends £12.99 on a trowel, £4.95 on flower pots and £1.62 on a bag of seeds.

Will Matt be given a free watering can? Explain your answer.

....

....

Multiplying and Dividing Decimals

You can multiply and divide decimals in exactly the same way as whole numbers.

EXAMPLE 1:

Lizzie makes 1.2 kg of fudge. She wants to give an equal amount to three friends. How much fudge should each friend get? Give your answer in kg.

Answer: divide the amount of fudge by the number of friends.

1.2 ÷ 3 = **0.4 kg**

EXAMPLE 2:

Chris is working out his petrol expenses.

He has driven 63.5 miles on business this week.
He is allowed to claim £0.30 per mile for petrol.

How much money can Chris claim for petrol this week?
Give your answer in pounds (£).

Answer: multiply the number of miles by the cost per mile.

63.5 × 0.30 = **£19.05**

Rounding off Decimals

1) You can sometimes get an answer with lots of numbers after the decimal point.
2) Instead of writing down the whole thing, you can shorten the answer and only write down one or two numbers after the decimal point. This is called rounding off.
3) To round off you need to decide how many numbers you want after the decimal point. Then look at the next number along (this number is called the decider).
4) If the decider is less than five you can just leave it off (and all the numbers after it) when you write down your answer.

EXAMPLES:

1) Round 2.8427865 so that there are two numbers after the decimal point.

   You want two numbers after the decimal point, so the decider is the third number after the decimal point. The decider is 2, which is less than 5.

   So the answer is **2.84**

2) Round 10.341346786 so that there is one number after the decimal point.

   You want one number after the decimal point, so the decider is the second number after the decimal point. The decider is 4, which is less than 5.

   So the answer is **10.3**

5) If the decider is 5 or more, then you need to add 1 to the last number when you round off.

EXAMPLES:

1) Round 9.3186895 so that there are two numbers after the decimal point.

   You want two numbers after the decimal point, so the decider is the third number after the decimal point. The decider is 8, which is more than 5, so you need to add 1 to the last number.

   So the answer is **9.32**

2) Round 20.85373122 so that there is one number after the decimal point. You want one number after the decimal point, so the decider is the second number after the decimal point. The decider is 5, so you need to add 1 to the last number.

   So the answer is **20.9**

Practice Questions

1) Round 3.57896 so that there is one number after the decimal point.

   ....

2) Round 1.024 so that there are two numbers after the decimal point.

   ....

3) Kim runs 5.3 km, three times a week. How far does she run in one week?

   ....

4) Sanjay does a sponsored walk. He raises £363.50. He wants to divide the money equally between his two favourite charities. How much money will each charity get?

   ....

5) Chris is painting his garden fence. He needs 0.6 litres of paint to cover one fence panel. How many litres of paint will he need to cover six fence panels?

   ....

6) Zoe is working out her petrol expenses. She drove a total of 104.2 km on a business trip. She can claim £0.40 per km for petrol. How much money is Zoe allowed to claim for petrol?

   ....

Percentages

Understanding Percentages

1) 'Per cent' means 'out of 100'.
2) % is a short way of writing 'per cent'.
3) So 20% means twenty per cent. This is the same as 20 out of 100.
4) You can write any percentage as a fraction. There's more on fractions on page 11.

$20\% = \frac{20}{100}$

Put the percentage on the top of the fraction.

Put 100 on the bottom of the fraction.

Calculating Percentages

1) Sometimes, you might need to calculate the 'percentage of' something.
2) In these cases, 'of' means 'times' (multiply).

EXAMPLE 1:

What is 20% of £60?

1) Write it down: 20% of £60 ↓ ↓ ↓
2) Turn it into maths: $\frac{20}{100} \times 60$
3) Type it into your calculator: 20 ÷ 100 × 60 = **£12**

EXAMPLE 2:

A jacket costs £42.

How much money could be saved with a voucher for 15% off?

The question wording is a bit trickier, but it's just asking you to find 15% of £42.

1) Write it down: 15% of £42 ↓ ↓ ↓
2) Turn it into maths: $\frac{15}{100} \times 42$
3) Type it into your calculator: 15 ÷ 100 × 42 = **£6.30**

Practice Questions

1) Write 34% as a fraction.

....

2) What is 12% of 40?

....

3) What is 75% of 15?

....

4) Kim teaches an aerobics class. The class has twenty members.

a) 80% of the class are women. How many members of the class are women?

....

b) 65% of the class are over 30. How many members of the class are over 30?

....

5) A dress costs £35. How much could be saved with a voucher for 10% off?

....

Calculating Percentage Increase

1) Sometimes, you might need to calculate a percentage increase.

2) If so, you need to find the 'percentage of' first.
Then you add it on to the original number.

EXAMPLE:

Cara earns 5% interest on her savings. She has £50 in her account.
How much money will she have once the interest has been added?

Answer:

Find 5% of £50: $\frac{5}{100} \times 50 = 5 \div 100 \times 50 = £2.50$

Add this on to £50: 50 + 2.50 = **£52.50**

Calculating Percentage Decrease

1) You might also need to calculate a percentage decrease.
2) First you find the 'percentage of'. Then you take it away from the original number.

EXAMPLE:

In the sale, a jacket which usually costs £60 has a 10% discount. What is the reduced price of the jacket?

Answer:

1) Find 10% of £60: $\frac{10}{100} \times 60 = 10 \div 100 \times 60 = £6$

2) Take this away from £60: $60 - 6 =$ **£54**

Practice Questions

1) Steve is buying a car. It would normally cost £3200, but today there is 20% off.

   a) How much money could Steve save today?

   ....

   b) What is the reduced price of the car?

   ....

2) Anja is buying material. She measures out 2.5 m. The shop owner gives her an extra 10% for free. How much material does Anja end up with?

   ....

   ....

3) David borrows £5000 to start up a business. He pays back the loan in one year, plus 9% interest. How much money does David pay back in total?

   ....

   ....

Fractions, Decimals and Percentages

These Fractions, Decimals and Percentages Are All the Same

The following fractions, decimals and percentages all mean the same thing.

They're really common, so it's a good idea to learn them.

$\frac{1}{2}$ is the same as 0.5, which is the same as 50%.

$\frac{1}{4}$ is the same as 0.25, which is the same as 25%.

$\frac{3}{4}$ is the same as 0.75, which is the same as 75%.

$\frac{1}{1}$ is the same as 1, which is the same as 100%.

You Can Change Fractions into Percentages

To change a fraction into a percentage you should:

1) Multiply the fraction by 100. 2) Add a per cent (%) sign.

EXAMPLE 1:

What is $\frac{2}{5}$ as a percentage?

1) Multiply the fraction by 100: $2 \div 5 \times 100 = 40$

2) Add a % sign = **40%**

EXAMPLE 2:

Last year Frank's horse won 3 of the 10 races it entered.
What percentage of its races did Frank's horse win?

Frank's horse won 3 out of 10 races, so the fraction is: $\frac{3}{10}$

1) Multiply the fraction by 100: $3 \div 10 \times 100 = 30$

2) Add a % sign = **30%**

You Can Also Convert Fractions into Decimals

To convert a fraction into a decimal you should:

Divide the top number in the fraction by the bottom number.

What is $\frac{2}{5}$ as a decimal?

Answer: divide 2 by 5. $2 \div 5 =$ **0.4**

Practice Questions

1) What is $\frac{1}{2}$ as a decimal?

....

2) A sale is offering a discount of 25%. What is this as a fraction?

....

3) Lyn buys $1\frac{1}{4}$ kg of cheese. Write $1\frac{1}{4}$ kg as a decimal.

....

4) What is $\frac{3}{5}$ as:

   a) a percentage? ....

   b) a decimal? ....

5) Cumbria County Council sends out a survey. 4 out of 10 people respond.

   a) What percentage is this?

   ....

   b) Lancashire County Council sends out a similar survey. 2 out 5 of people respond. Which council has a higher percentage of people responding to their survey?

   ....

   ....

Comparing Fractions, Percentages and Decimals

You need to be able to compare fractions, percentages and decimals.

EXAMPLE 1:

Which is greater, 0.5 or $\frac{6}{10}$?

You need to work out what $\frac{6}{10}$ is as a decimal.

To convert $\frac{6}{10}$ to a decimal, divide 6 by 10: $6 \div 10 = 0.6$

0.6 is bigger than 0.5, so $\mathbf{\frac{6}{10}}$ **is greater**.

EXAMPLE 2:

Tony is booking a holiday. The travel agent offers Tony two deals:

"All flights half price" OR "25% off all hotels"

The flights Tony wants to book normally cost £300.
The hotel he wants to book normally costs £400.

Which offer will save Tony the most money?

First work out how much money Tony will save on the flights: $\frac{1}{2} \times 300 = £150$

Then work out how much he'll save on the hotel:

25% of 400 = $\frac{25}{100} \times 400 = £100$

The **half price flights** offer will save Tony the most money.

Practice Questions

1) Which is greater, 0.25 or $\frac{2}{10}$?

....

2) Leanne buys a new TV in the sale. It would normally cost £500, but she gets 20% off.
Darren also buys a new TV in the sale. It would normally cost £540, but he gets a third off.

Who ends up paying less for a TV, Leanne or Darren? Explain your answer.

....

....

....

Ratios

Ratios Compare One Part to Another Part

Ratios are a way of showing how many things of one type there are compared to another.

Look at this pattern:

There are two white tiles and six blue tiles.
In other words, for every white tile there are three blue tiles.

So the ratio of blue tiles to white tiles is 3:1.

The order the numbers are written in the ratio depends on the order of the words — the ratio of blue tiles to white tiles is 3:1. The ratio of white tiles to blue tiles is 1:3.

Questions Involving Ratios

To answer a question involving ratios, you usually need to start by working out the value of one part. For example, the cost of one thing or the mass of one part.

You can then use this to answer the question.

EXAMPLE 1:

5 pints of milk cost £2.00. How much will 3 pints cost?

1) First, you need to find out how much 1 pint of milk costs.
   You know that 5 pints cost £2, so you need to divide £2 by 5.

   cost of 1 pint = 2 ÷ 5 = 0.4

2) To work out the cost of 3 pints, times your answer by 3.

   0.4 × 3 = **£1.20**

EXAMPLE 2:

A drink is made from 1 part cordial and 3 parts water.
800 ml of the drink is made. How much cordial is used?

1) First you need to work out how many parts there are in total.
   To do this, add up the numbers in the ratio.

   1 + 3 = 4 parts

2) The drink contains 1 part cordial. To work out how many ml are in 1 part, divide the total amount by the number of parts:

   800 ÷ 4 = **200 ml**

EXAMPLE 3:

£5000 will be split between two people in the ratio 1:4.
How much money does each person get?

1) First, work out how many parts the £5000 will be split into in total.
   To do this, add up the numbers in the ratio.

   $1 + 4 = 5$ parts

2) To find out how much one part is worth, divide 5000 by 5: $5000 \div 5 = 1000$

3) The first person in the ratio gets one part, so they get **£1000**.

4) The second person in the ratio gets four parts.
   To work out how much money they get, times the value of one part by 4:

   $1000 \times 4 =$ **£4000**

Practice Questions

1) Helen is making orange drink. She mixes 4 parts water to 1 part squash.

   a) What is the ratio of squash to water in Helen's drink?

   ....

   b) Helen wants to make 500 ml of orange drink. How much squash does she need?

   ....

   ....

2) Jake is tiling his bathroom floor. He uses three green tiles for every white tile.
   Jake uses twenty four tiles in total. How many of them are green?

   ....

   ....

3) £3000 will be split between two people in the ratio 2:1. How much does each person get?

   ....

   ....

   ....

Working Out Total Amounts

1) You can use ratios to work out total amounts.
2) The first step is usually to work out the total number of parts.
3) The second step is usually multiplying the total number of parts by the value of one part.

EXAMPLE:

A drink is made from one part cordial and three parts water. 70 ml of cordial is used. How much drink is made?

1) Find the total number of parts for the drink.
   To do this, add up the numbers in the ratio: 1 + 3 = 4
2) Times the total number of parts by the amount given for one part: 4 × 70 = **280 ml**

One part cordial = 70 ml

Scaling Recipes Up or Down

Lots of recipes use ratios. Ratios stay the same even when amounts change.

EXAMPLE 1:

Amy is making brownies. The recipe says to use 6 pieces of chocolate for every 2 eggs. Amy is using 10 eggs.

How many pieces of chocolate does she need?

Answer: 10 eggs is five times as many eggs as in the recipe.

So Amy will need five times as many pieces of chocolate to match.

6 × 5 = **30 pieces** of chocolate

EXAMPLE 2:

Nick is making lasagne. His recipe says to use 1 tin of tomatoes for every 500 g of mince. The recipe serves 4 people.

Nick wants to make lasagne for 12 people.
How much mince will he need to use?

Answer: 12 people is three times as many as the recipe serves.
So Nick needs to make three times as much lasagne.

500 × 3 = **1500 g** of mince

Practice Questions

1) Hannah is mixing concrete. She mixes 1 part cement to 3 parts sand.
   She uses 12 kg of cement. How much concrete will she have in total?

   ....

   ....

2) Aled is thinning some paint. He mixes 1 part paint thinner to 4 parts paint.
   He uses 150 ml of paint thinner. How much thinned paint will he end up with?

   ....

   ....

3) Neil is making fairy cakes. The recipe says to use 120 g of flour for every 2 eggs.
   The recipe makes 12 cakes.

   a) Neil uses 240 g of flour. How many eggs does he need to use?

      ....

      ....

   b) How much flour would Neil need to make 36 cakes?

      ....

      ....

4) Gina and Clive inherit some money. The money is split in a 1:2 ratio, with Clive getting the most money. Gina inherits £500. How much money did the pair inherit in total?

   ....

   ....

Formulas in Words

A Formula is a Type of Rule

1) A formula is a rule for working out an amount.
2) Formulas can be written in words. Sometimes, it can be tricky to spot the formula.

EXAMPLE:

Sam is paid £5.20 per hour. How much does he earn in 8 hours?

You're told that: "Sam is paid £5.20 per hour." This is a formula.
You can use it to work out how much Sam earns in a given number of hours.

1) The calculation you need to do here is:

   Sam's pay = 5.20 × number of hours

2) You've been asked how much Sam earns in 8 hours, so put '8' into the calculation in place of 'number of hours':

   5.20 × 8 = **£41.60**

You can use the same formula to work out how much Sam earns for any number of hours.

Formulas Can Have More Than One Step

Some formulas have two steps in them. You need to be able to use two-step formulas.

EXAMPLE:

A ham takes 30 minutes per kilogram to cook, plus an extra 25 minutes.
How long does a 1.5 kg ham take to cook?

The formula here is "30 minutes per kilogram, plus 25 minutes".

1) Work out the calculation you need to do:

   Step 1 = 30 × number of kilograms

   Step 2 = + 25

   Cooking time = (30 × number of kilograms) + 25

   There's more on brackets on page 6.

2) Then just stick the right numbers in.
   In this case it's '1.5' in place of 'number of kilograms':

   (30 × 1.5) + 25 = **70 minutes**

Function Machines

Function machines can help you to use formulas with more than one step.

EXAMPLE 1:

This function machine helps you to work out the cooking time for a ham.
It gives you the cooking time of the ham in minutes.

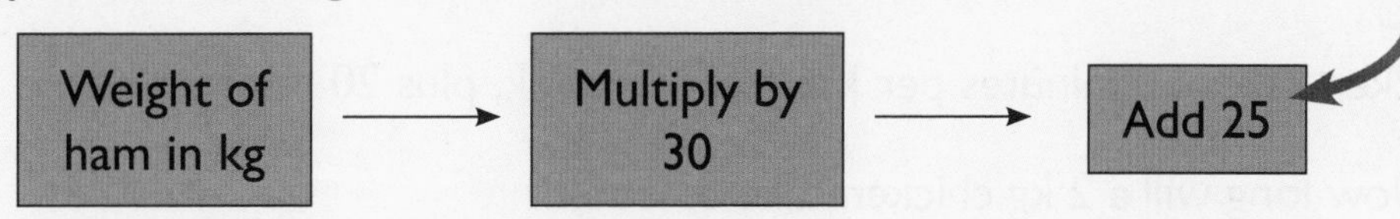

To work out how long a 1.5 kg ham takes to cook:

1) Put 1.5 into the function machine in place of 'Weight of ham in kg'.

2) Follow the rest of the steps.

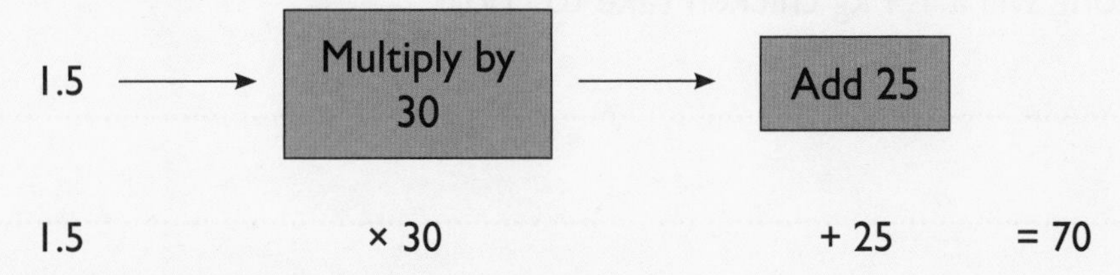

So the ham takes **70 minutes** to cook.

EXAMPLE 2:

Carrie is checking her electricity bill.
To work out how much her electricity costs, she uses this rule:

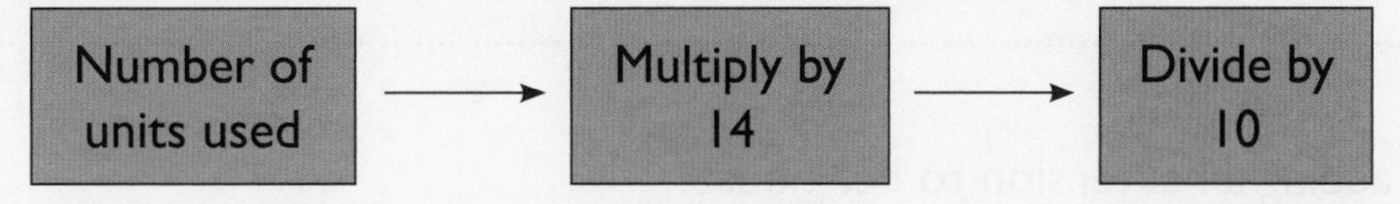

This gives her the cost of the electricity she has used in pounds (£).

Carrie has used 154 units of electricity so far this year.
How much should she be charged?

1) Put 154 into the function machine, in place of 'Number of units used'.

2) Follow the rest of the steps.

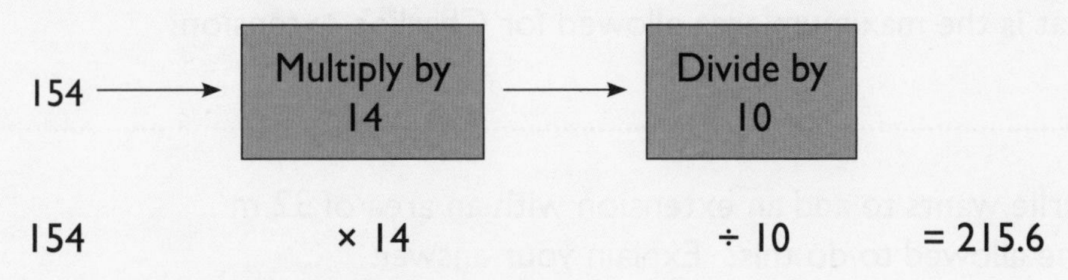

So Carrie will be charged **£215.60.**

Practice Questions

1) Gary earns £7.50 per hour. How much will Gary earn in 7.5 hours?

2) A chicken takes 50 minutes per kilogram to cook, plus 20 minutes extra.

   a) How long will a 2 kg chicken take to cook?

   b) How long will a 1.4 kg chicken take to cook?

3) A broadband deal is £10 per month for 3 months, then £16 per month after that.
   How much will 12 months' broadband cost?

4) Charlie is adding an extension to her house.
   She wants to know the maximum area that the extension is allowed to cover.
   When she asks the council, they tell her to work it out using this rule:

   Area of the house (in m²) → Divide by 4 → Add 10

   This will give Charlie the maximum area allowed for the extension in m².

   The area of Charlie's house is 90 m².

   a) What is the maximum area allowed for Charlie's extension?

   b) Charlie wants to add an extension with an area of 32 m².
      Is she allowed to do this? Explain your answer.

Units

All Measures Have Units

1) Almost everything that you measure has units. For example, metres (m) or grams (g).
2) They're really important. For example, you can't just say that a distance is 4 — you need to know if it's 4 miles, 4 metres, 4 kilometres, etc.

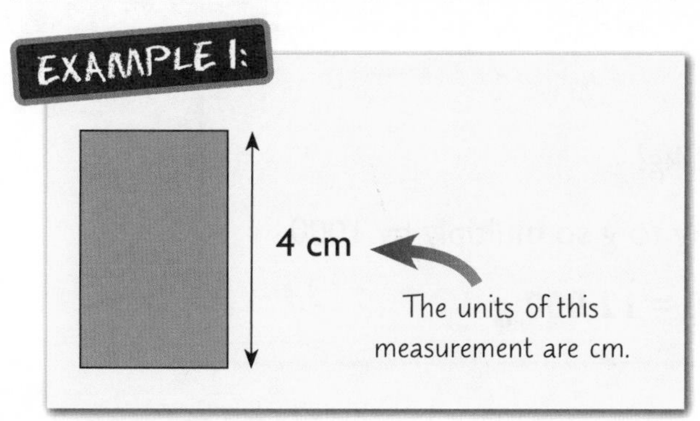

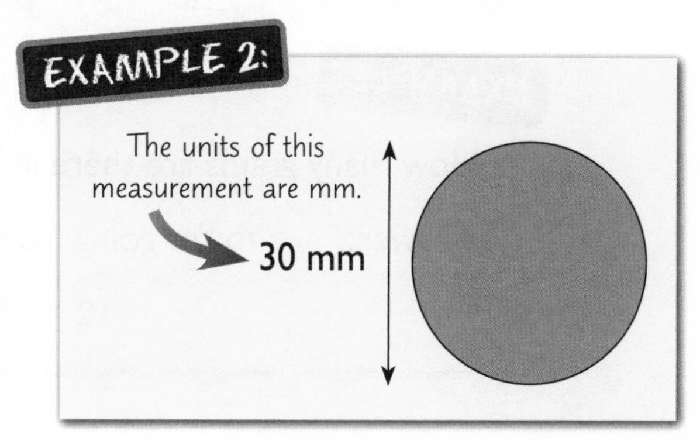

Units of Length

1) Length is how long something is. Some common units for length are millimetres (mm), centimetres (cm), metres (m) and kilometres (km).
2) Here's how some of these units are related:
3) Sometimes you might need to change something from one unit to another.
4) To switch between mm, cm, m and km you can multiply or divide by 10, 100, or 1000.

Length
1 cm = 10 mm
1 m = 100 cm
1 km = 1000 m

To go from mm to cm, divide by 10.
To go from cm to m, divide by 100.
To go from m to km, divide by 1000.

To go from cm to mm, multiply by 10.
To go from m to cm, multiply by 100.
To go from km to m, multiply by 1000.

EXAMPLES:

How many millimetres are there in 20 cm?

Answer: You're going from cm to mm, so multiply by 10.

20 × 10 = **200 mm**

What is 3500 m in km?

Answer: You're going from m to km, so divide by 1000.

3500 ÷ 1000 = **3.5 km**

Units of Weight

1) Weight is how heavy something is. Grams (g) and kilograms (kg) are common units for weight.
2) Here's how to change between g and kg...

Weight
1 kg = 1000 g

To go from g to kg, divide by 1000.

To go from kg to g, multiply by 1000.

EXAMPLE:

How many grams are there in 12 kg?

Answer: You're going from kg to g so multiply by 1000.

$12 \times 1000 =$ **12 000 g**

Units of Capacity

1) Capacity is how much something will hold. Common units are millilitres (ml), centilitres (cl) and litres (L) .
2) To change between ml, cl and L you can multiply or divide by 10 or 100.

Capacity
1 cl = 10 ml
1 L = 100 cl

To go from ml to cl, divide by 10.

To go from cl to L, divide by 100.

To go from cl to ml, multiply by 10.

To go from L to cl, multiply by 100.

EXAMPLE:

How many centilitres are in 200 ml?

Answer: You're going from ml to cl, so divide by 10.

$200 \div 10 =$ **20 cl.**

Practice Questions

1) How many cm are in 1 m?

....

2) Maria needs to weigh out 1300 g of flour. Write 1300 g in kg.

....

3) Imran needs to use 5 cl of wine to make a stew. His measuring jug only measures in ml. How many millilitres of wine should he use?

....

4) Ivy is planning to swim 5000 m for charity. How far is she planning to swim in kilometres?

....

Using Conversion Factors

1) Sometimes you can't change from one unit to another by multiplying or dividing by 10, 100 or 1000.
2) You may need to multiply or divide by a different number instead (you'll be given the number in the question somewhere). This number is called the conversion factor.
3) If you're not sure whether to multiply or divide, then do both and pick the common-sense answer — think about whether the smaller or bigger answer is sensible.

EXAMPLE 1:

If 1 metre is equal to 3.28 feet, how many feet long is a 5 metre van?

1) First find the conversion factor — here it's 3.28.
2) Multiply or divide by the conversion factor, or do both if you're not sure. → $5 \times 3.28 = 16.4$; $5 \div 3.28 = 1.52$
3) The answer must be 16.4 feet or 1.52 feet.
   If 1 metre is about 3 feet, then 5 metres must be bigger than 1.52 feet.

   So the van must be **16.4 feet** long.

EXAMPLE 2:

Myles can run at 7 miles per hour.
1 mile per hour is the same as 1.6 km per hour.

How fast can Myles run in km per hour?

1) The conversion factor is 1.6.
2) Multiply or divide by the conversion factor, or do both if you're not sure. → $7 \times 1.6 = 11.2$
   $7 \div 1.6 = 4.4$
3) 1 mile per hour is 1.6 km per hour, so Myles' speed in km per hour must be higher than his speed in miles per hour.

   So, Myles can run at **11.2 km** per hour.

Practice Questions

1) 1 inch is equal to 2.5 cm. How many centimetres long is a 5 inch sausage roll?

   ....

   ....

2) 1 kilogram is equal to 2.2 pounds. How many kilograms is a 5.6 pound weight?

   ....

   ....

3) Tyrone's Post Office® will give him £0.80 for every euro he has.
   Tyrone has 14 Euros. How many pounds will the Post Office® give Tyrone?

   ....

   ....

4) Alan wants to know how many Calories are in his sandwich.
   The packaging says that the sandwich contains 1800 kJ of energy.
   1 kJ is equal to 0.2 Calories. How many calories does his sandwich contain?

   ....

   ....

Weight

Weight is How Heavy Something is

1) Weight can be measured in lots of different units. For example, grams (g), kilograms (kg), ounces (oz), pounds (lb) and stones (st).

2) You need to be able to solve problems involving weight.

EXAMPLE 1:

Arron has to take 5 boxes in a lift. The lift can carry 500 kg at a time. Each box weighs 100 kg. Arron weighs 90 kg. How many trips will he have to take?

1) Work out the total weight of the boxes. → 5 × 100 kg = 500 kg

2) Work out the weight of the boxes plus Arron. → 500 kg + 90 kg = 590 kg

The lift can only carry 500 kg, so he'll need to take more than one trip.

3) Work out how many trips he needs to take.

On the first trip he can take himself and 4 boxes: 4 × 100 kg + 90 kg = 490 kg

On the second trip he can take himself and 1 box: 100 kg + 90 kg = 190 kg

So, it will take Arron **two trips** to take all of the boxes in the lift.

EXAMPLE 2:

Roan has a bad back. His doctor told him not to lift more than 3000 g at a time. Roan has bought one bag of flour, two bags of rice and four chocolate bars.

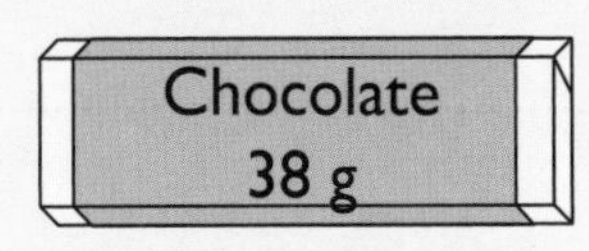

Can Roan carry all of his shopping back safely?

1) First you need everything in the same units, so change the weight of the flour into grams. → 1.5 kg × 1000 = 1500 g.

2) Next work out what weight of rice and chocolate he has.

Don't forget — he's bought 2 bags of rice and 4 chocolate bars. → Rice: 500 g × 2 = 1000 g

Chocolate: 38 g × 4 = 152 g

3) Then work out the total weight. → 1500 g + 1000 g + 152 g = 2652 g

2652 g is less than 3000 g, so Roan **can** carry his shopping back safely.

Practice Questions

1) Add up the following weights: 142 g, 263 g, 657 g, 4 g.

....

2) Amera used to weigh 108 kg. She has lost 3 kg. What does she weigh now?

....

3) Ewa is designing a child's chair. The chair needs to be able to carry a weight of 90 kg. How many 10 kg weights will Ewa need to test the strength of the chair?

....

....

4) David is a jockey. To compete in the local horse race, he and his equipment have to weigh less than 57 kg. David weighs 51 kg and his equipment weighs 5400 g. Can David take part in the race? Explain your answer.

....

....

5) Washing powder comes in three different sized boxes, 1 kg, 2.5 kg and 5 kg. Finlay wants to buy exactly 14 kg of washing powder. What is the smallest number of boxes Finlay can buy?

....

....

6) Samah has made 20 kg of chilli. There are 100 g of chilli in a portion. How many portions of chilli has Samah made?

....

....

7) Kane has to take five 5000 g parcels from the Post Office® to his van. Kane can safely lift 15 kg at once. How many trips will Kane need to take to move all of the parcels?

....

....

Capacity

Volume and Capacity

Volume is the amount of space something takes up.

Capacity is how much something will hold.

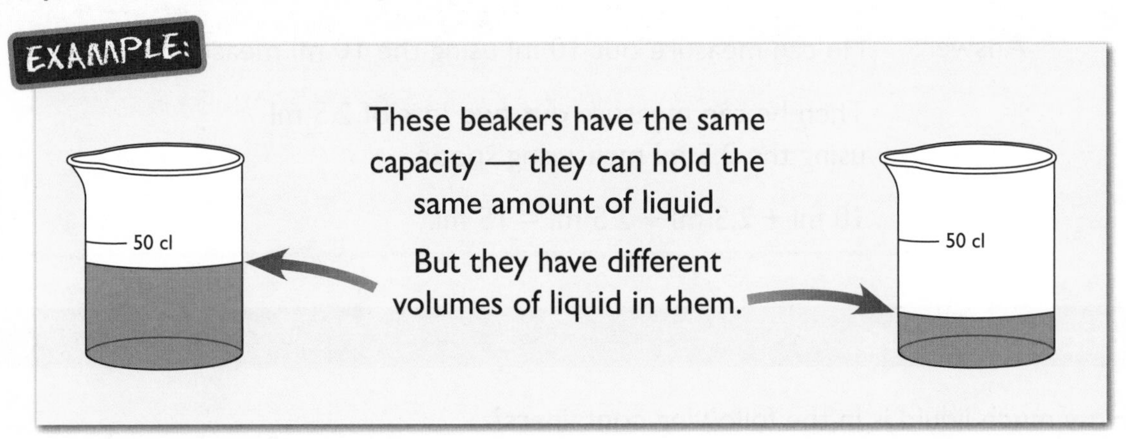

Reading Scales

A scale is something that you use to measure things
— like rulers, kitchen scales, measuring jugs and thermometers.

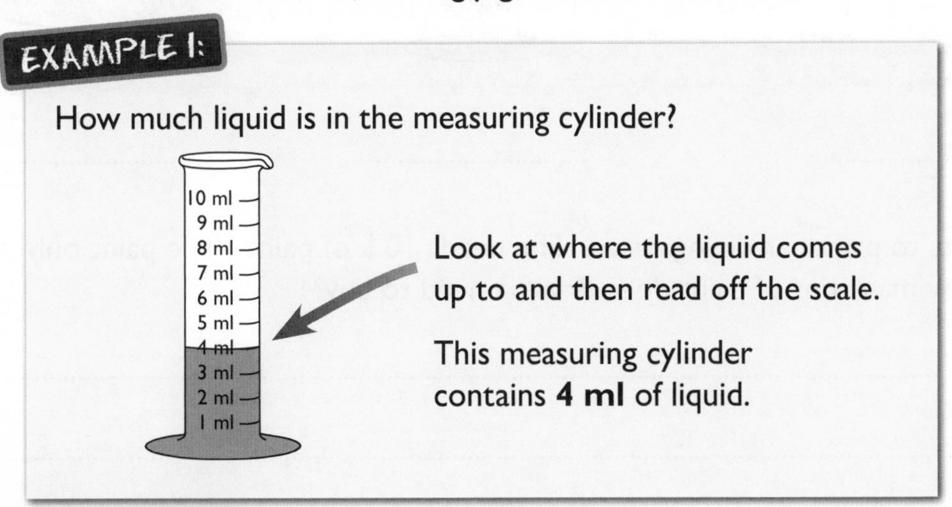

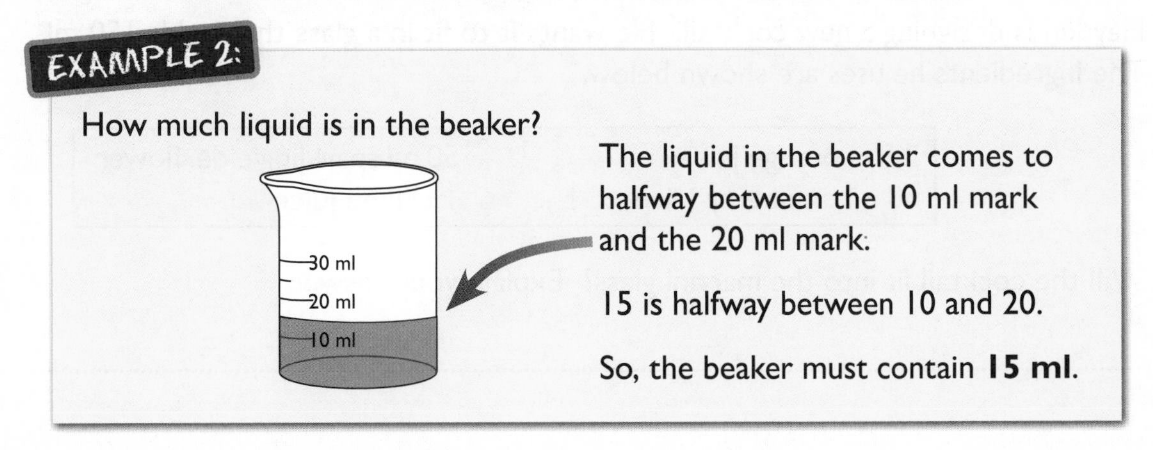

Questions Involving Capacity and Volume

You need to be able to solve problems involving capacity and volume.

EXAMPLE:

Andrew needs to measure out 15 ml of vanilla essence.
He has one 10 ml and one 2.5 ml measuring spoon.
How can Andrew measure out his vanilla essence?

Answer: He can measure out 10 ml using the 10 ml measuring spoon.

Then he can measure out two lots of 2.5 ml using the 2.5 ml measuring spoon.

10 ml + 2.5 ml + 2.5 ml = 15 ml.

Practice Questions

1) How much liquid is in the following containers?

a)

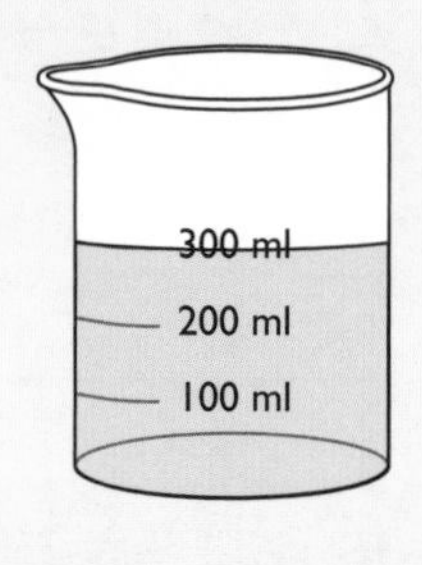

b)

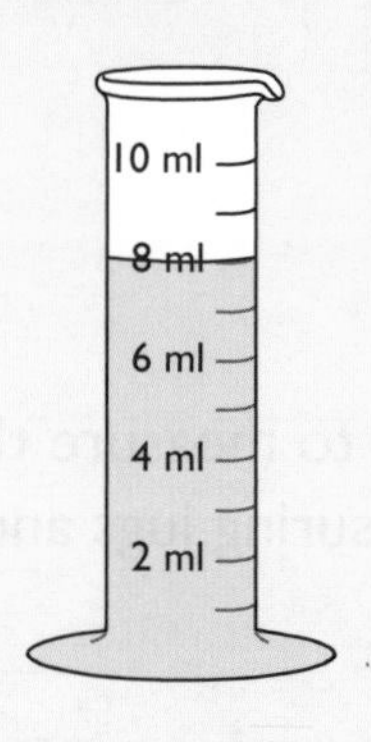

c)

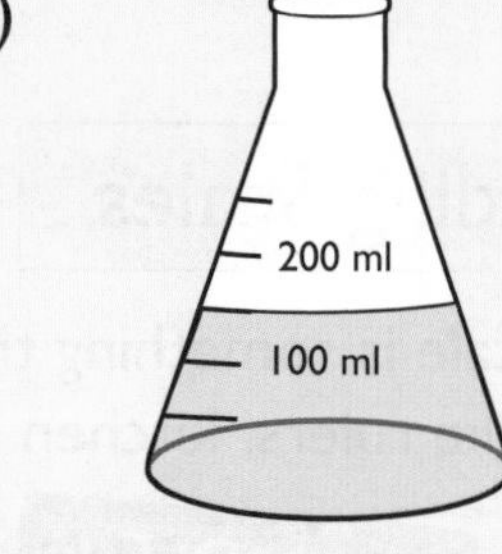

....

2) Brianna wants to paint her dining room. She needs 10 L of paint. The paint only comes in 2 L tins. How many tins of paint does Brianna need to buy?

....

....

3) Hayden is designing a new cocktail. He wants it to fit in a glass that holds 150 ml. The ingredients he uses are shown below.

2.5 cl orange juice	50 ml sparkling elderflower
50 ml cranberry juice	1 cl lime juice

Will the cocktail fit into the martini glass? Explain your answer.

....

Length and Perimeter

Finding the Perimeter of a Shape

The perimeter is the distance around the outside of a shape.

To find a perimeter, you add up the lengths of all the sides.

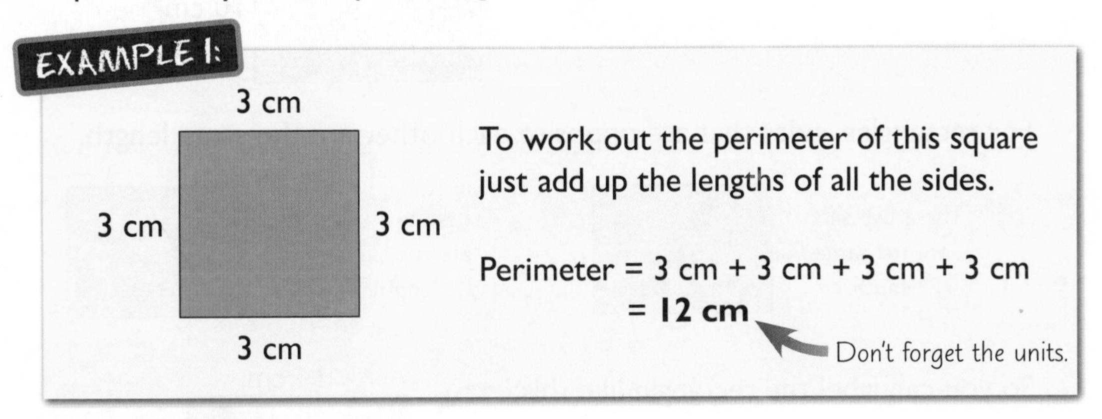

To work out the perimeter of this square just add up the lengths of all the sides.

Perimeter = 3 cm + 3 cm + 3 cm + 3 cm
= **12 cm**

Don't forget the units.

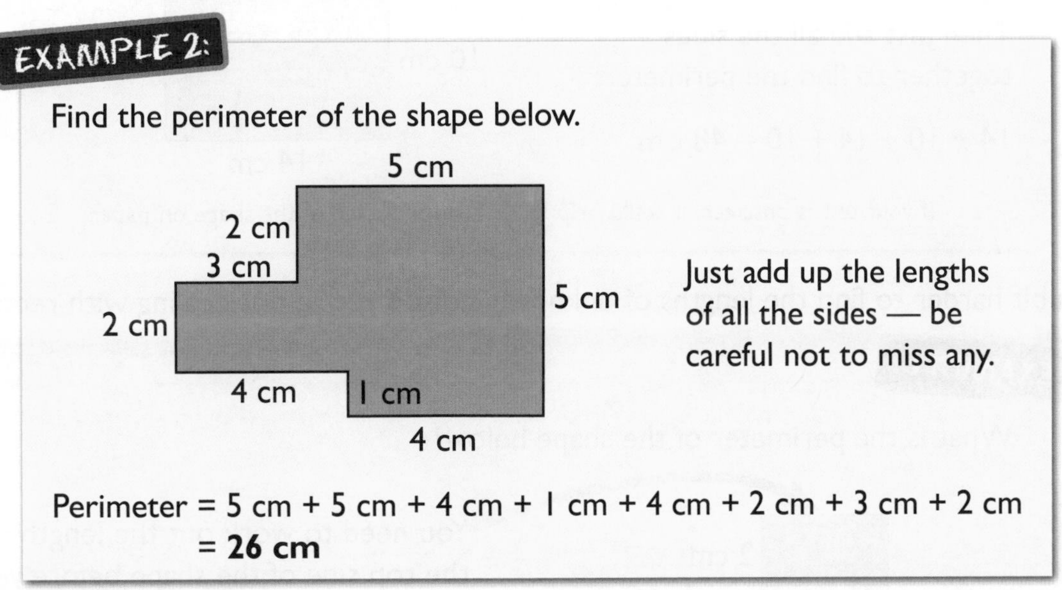

Find the perimeter of the shape below.

Just add up the lengths of all the sides — be careful not to miss any.

Perimeter = 5 cm + 5 cm + 4 cm + 1 cm + 4 cm + 2 cm + 3 cm + 2 cm
= **26 cm**

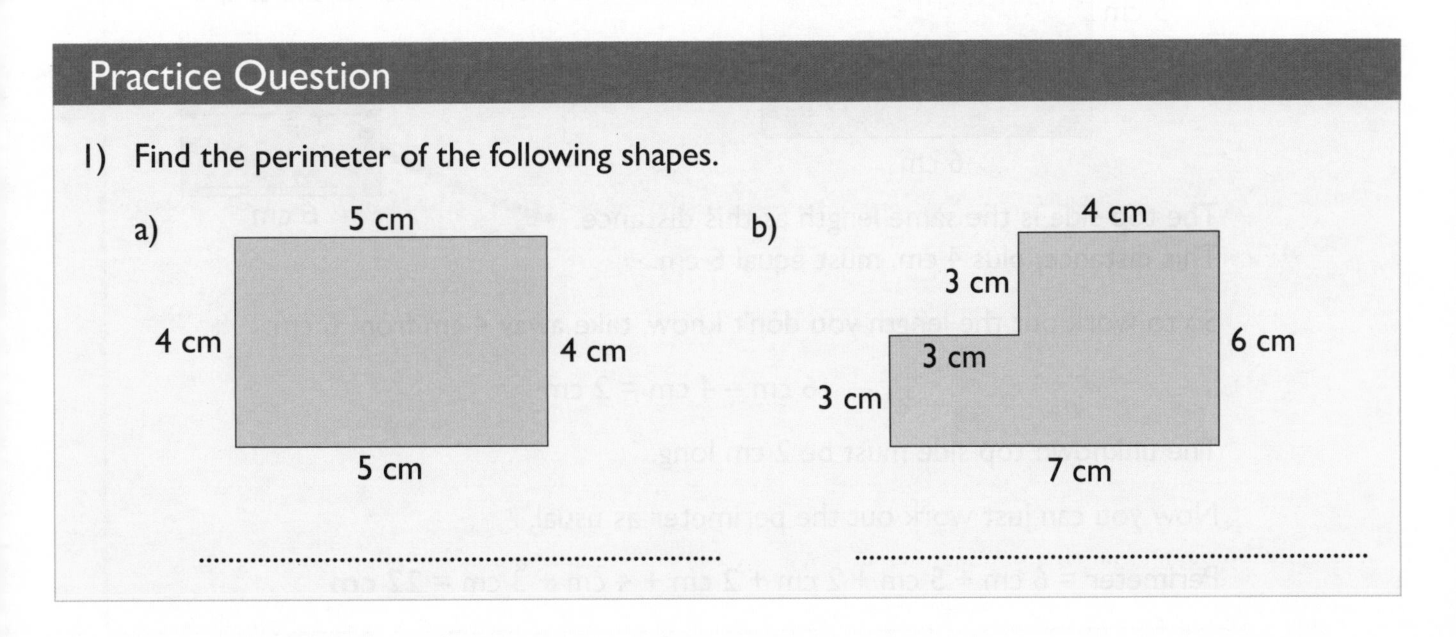

Practice Question

1) Find the perimeter of the following shapes.

a) ....

b) ....

Working Out the Length of an Unknown Side

If you're only given the lengths of some of the sides, you'll have to work out the rest before you can calculate the perimeter. Sometimes this is fairly simple.

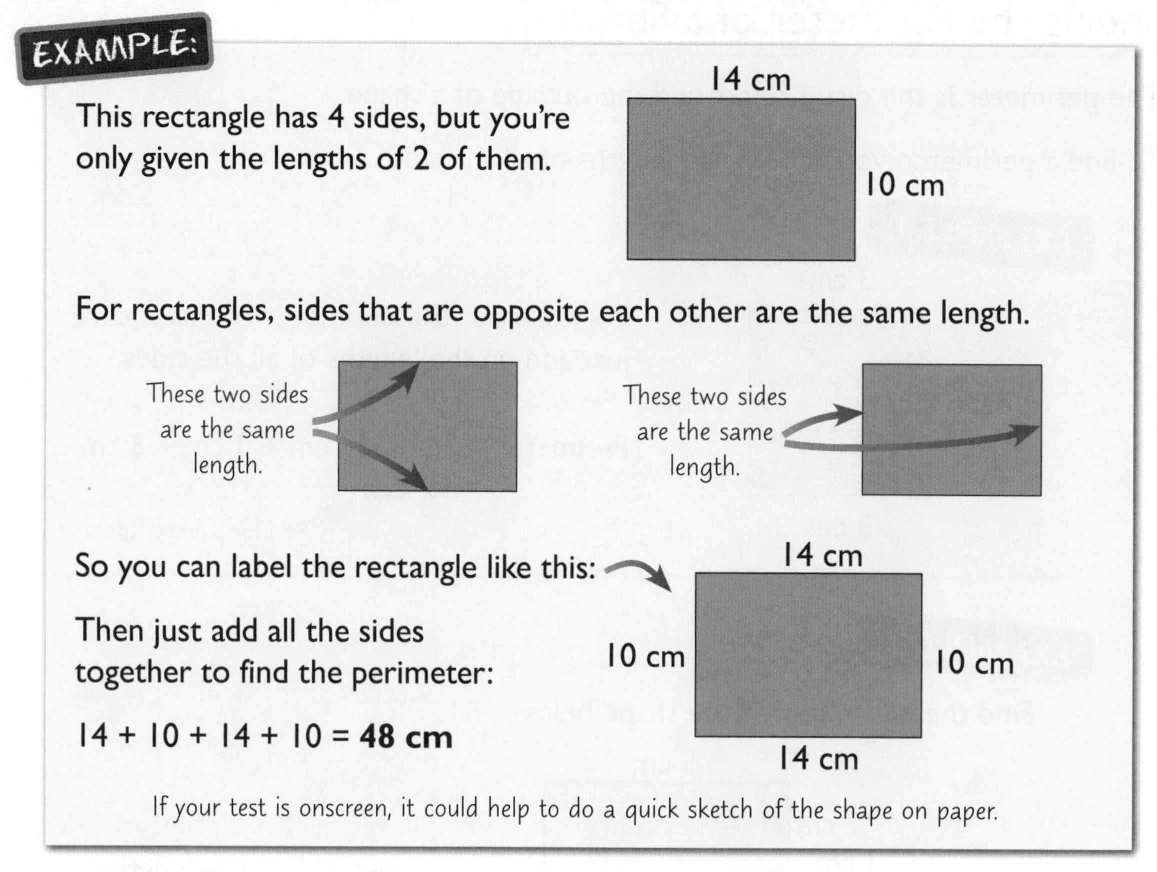
EXAMPLE:

This rectangle has 4 sides, but you're only given the lengths of 2 of them.

For rectangles, sides that are opposite each other are the same length.

So you can label the rectangle like this:

Then just add all the sides together to find the perimeter:

14 + 10 + 14 + 10 = **48 cm**

If your test is onscreen, it could help to do a quick sketch of the shape on paper.

It's a bit harder to find the lengths of unknown sides if you're not dealing with rectangles.

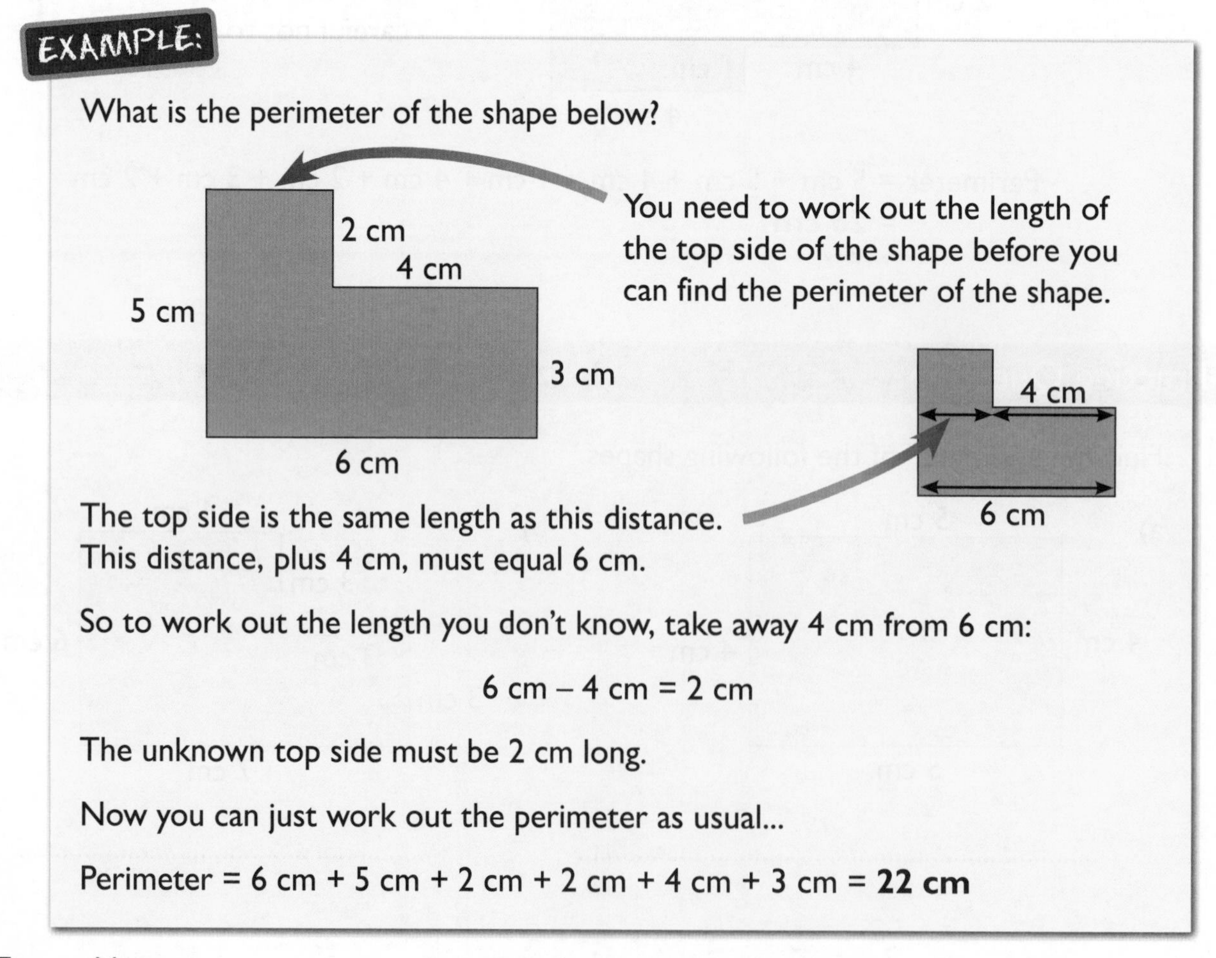
EXAMPLE:

What is the perimeter of the shape below?

You need to work out the length of the top side of the shape before you can find the perimeter of the shape.

The top side is the same length as this distance.
This distance, plus 4 cm, must equal 6 cm.

So to work out the length you don't know, take away 4 cm from 6 cm:

6 cm – 4 cm = 2 cm

The unknown top side must be 2 cm long.

Now you can just work out the perimeter as usual...

Perimeter = 6 cm + 5 cm + 2 cm + 2 cm + 4 cm + 3 cm = **22 cm**

Practice Questions

1) Calculate the perimeter of the following shapes.

a)

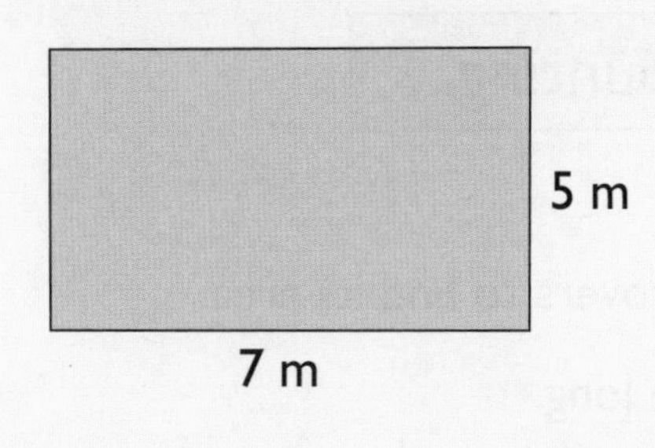

b)

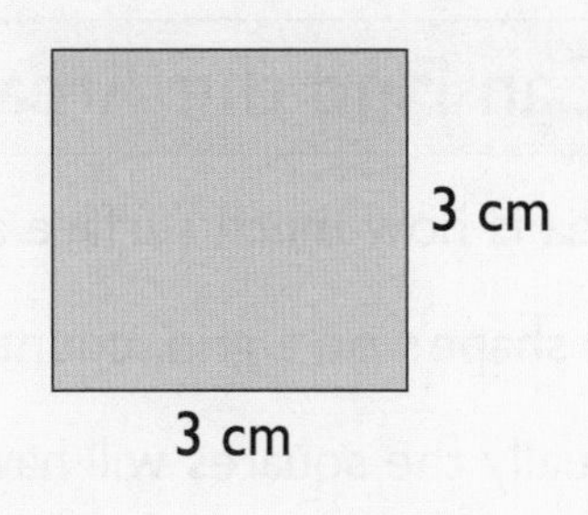

2) Calculate the perimeter of the following shapes.

a)

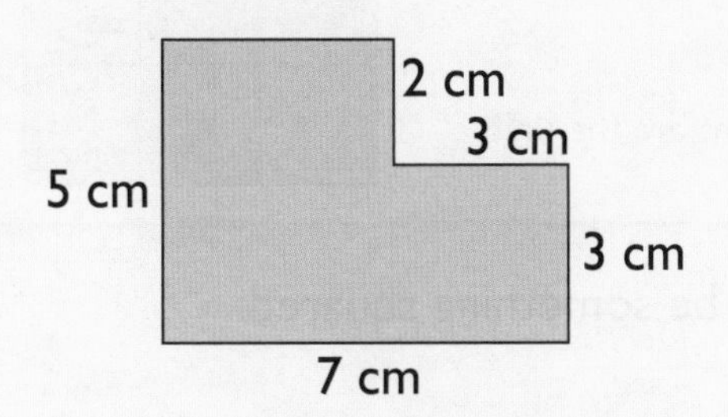

b)

8 cm

3 cm

6 cm

4 cm

c)

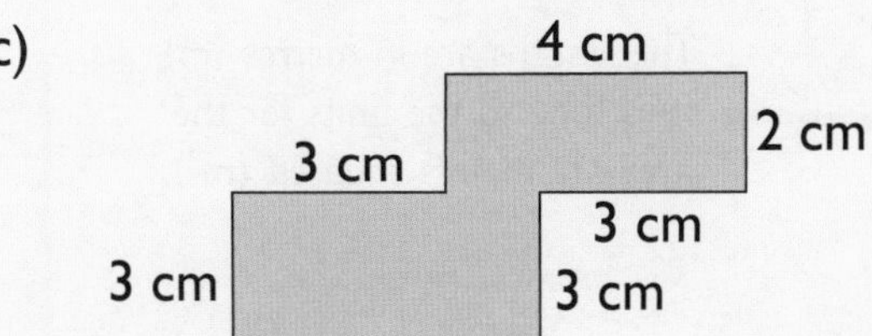

3) Quentin is fitting a skirting board in his living room. A sketch of his living room is shown below. The door to the room is 0.75 m wide and doesn't need skirting board attached to it. Calculate the length of skirting board that Quentin needs to buy.

4 m

3 m

4 m

Area

You Can Find the Area of Shapes by Counting...

1) Area is how much surface a shape covers.
2) If a shape's on a grid, count how many squares it covers to find its area.
3) Usually the squares will have sides one centimetre long.
   If so, each square is 1 centimetre squared. This is written as 1 cm^2.

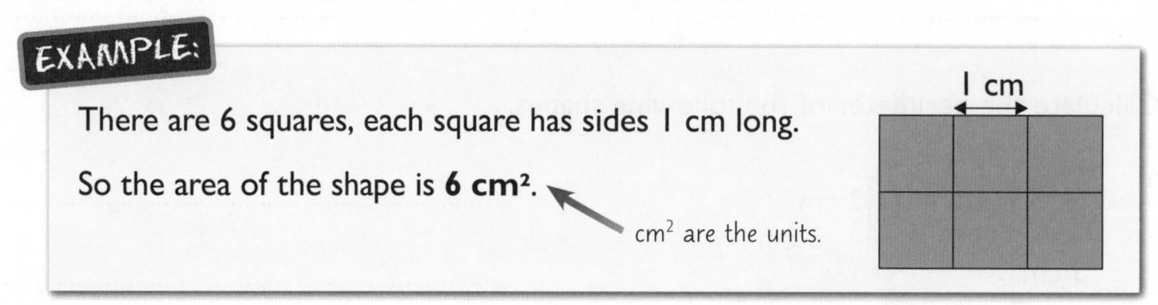

4) If you're dealing with area, the units will usually be something squared.
   For example, cm^2, m^2, mm^2.

...or by Multiplying

1) You can work out the area of rectangles by multiplying.
2) You need to know the lengths of the sides, then just multiply them together.

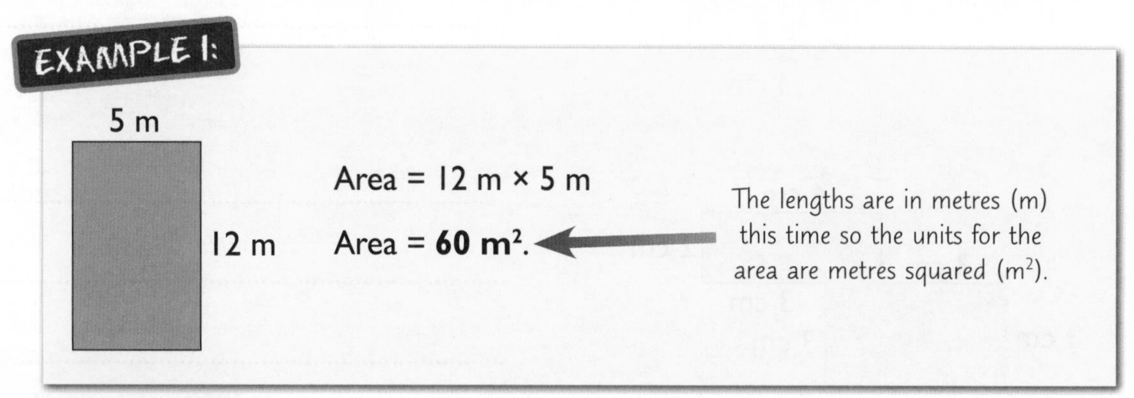

EXAMPLE 2:

Sophie is buying a new carpet for her dining room. The room is 6 m long and 5 m wide. What area of carpet does she need to buy?

Answer: 6 m × 5 m = **30 m^2**

6 m

5 m

You could draw a sketch of the room if it helps you.

Sometimes You Need to Split Shapes Up to Find the Area

It's a bit trickier to find the area of a shape that isn't a rectangle...

...but you can sometimes do it by splitting the shape up into rectangles.

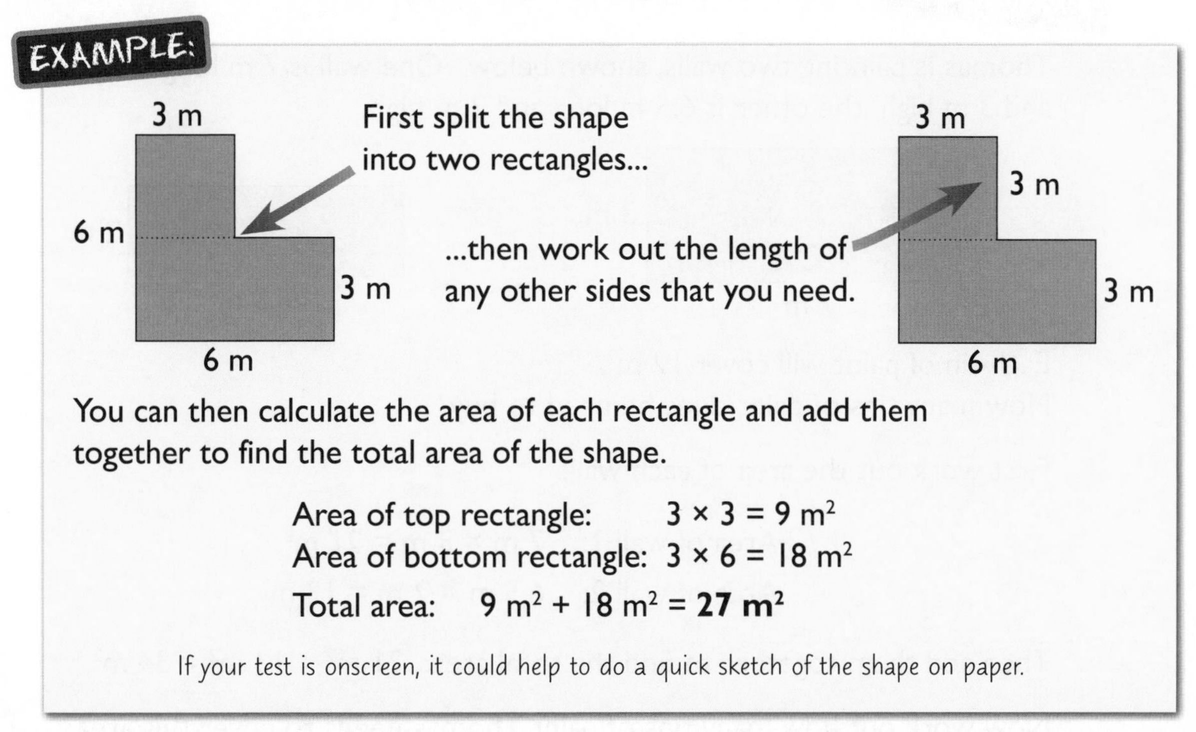

You can then calculate the area of each rectangle and add them together to find the total area of the shape.

Area of top rectangle: $3 \times 3 = 9\ m^2$
Area of bottom rectangle: $3 \times 6 = 18\ m^2$
Total area: $9\ m^2 + 18\ m^2 =$ **$27\ m^2$**

If your test is onscreen, it could help to do a quick sketch of the shape on paper.

Practice Question

1) Find the area of each of the shapes below:

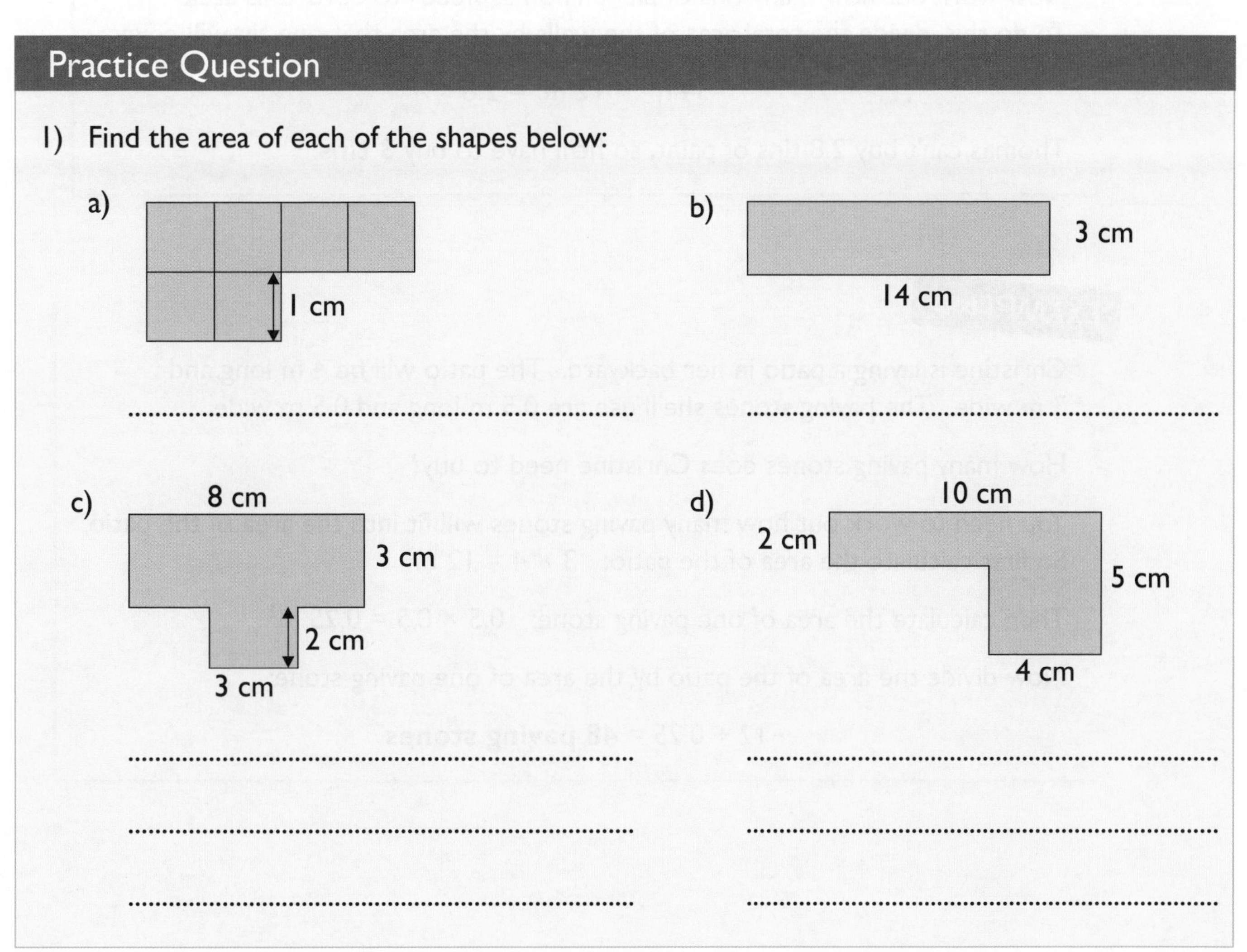

Using Areas in Calculations

Sometimes you'll need to work out an area as part of a bigger calculation.

EXAMPLE 1:

Thomas is painting two walls, shown below. One wall is 7 m long and 3 m high, the other is 6.5 m long and 2 m high.

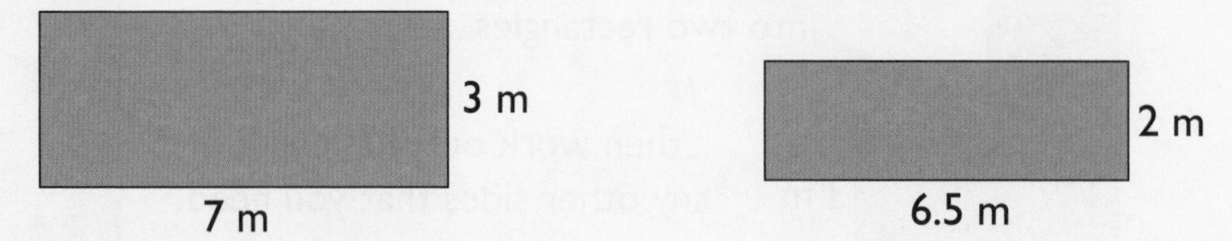

Each tin of paint will cover 12 m^2.
How many tins of paint does he need to buy?

First work out the area of each wall:

Area of wall 1: $7 \text{ m} \times 3 \text{ m} = 21 \text{ m}^2$
Area of wall 2: $6.5 \text{ m} \times 2 \text{ m} = 13 \text{ m}^2$

Then add them together to find the total area: $21 \text{ m}^2 + 13 \text{ m}^2 = 34 \text{ m}^2$

Now work out how many tins of paint Thomas needs to cover this area.
To do this, divide the total area of the walls by the area that one tin will cover:

$$34 \text{ m}^2 \div 12 \text{ m}^2 = 2.8$$

Thomas can't buy 2.8 tins of paint, so he'll have to buy **3 tins**.

EXAMPLE 2:

Christine is laying a patio in her backyard. The patio will be 4 m long and 3 m wide. The paving stones she'll use are 0.5 m long and 0.5 m wide.

How many paving stones does Christine need to buy?

You need to work out how many paving stones will fit into the area of the patio.
So first calculate the area of the patio: $3 \times 4 = 12 \text{ m}^2$

Then calculate the area of one paving stone: $0.5 \times 0.5 = 0.25 \text{ m}^2$

Now divide the area of the patio by the area of one paving stone:

$12 \div 0.25 =$ **48 paving stones**

Practice Questions

1) A church is raising money to put gravel on their car park and path. It will cost £4.20 for each m^2 that needs gravel. How much money does the church need to raise?

car park 10 m
15 m

path 1.5 m
10 m

2) Emma is tiling a section of wall in her bathroom. The tiles are 10 cm by 10 cm. The section of wall she wants to tile is 120 cm across and 80 cm high.

Section of wall 80 cm
120 cm

Tile 10 cm
10 cm

How many tiles will Emma need?

Volume

You Can Calculate the Volume of a Shape by Counting...

1) Volume is how much space something takes up.
2) Sometimes you can calculate the volume of a shape by counting cubes.
3) The units for volume will usually be something 'cubed'. This is shown by a ^{3} in the units. For example, cm^3 (centimetres cubed) or m^3 (metres cubed).

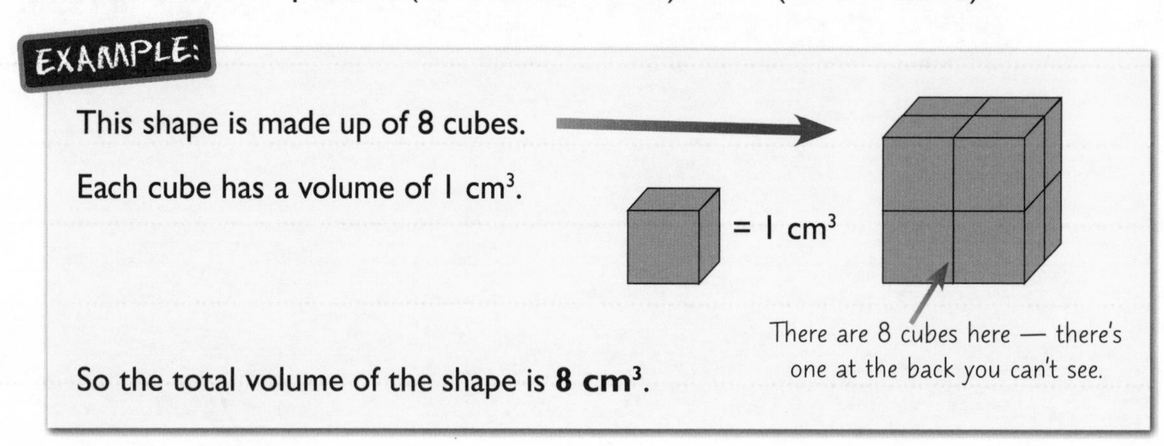

...or by Multiplying

1) You can work out the volume of shapes even if they aren't broken down into cubes.
2) For some shapes, you just need to know the length, the width and the height.
3) Then just multiply them together.

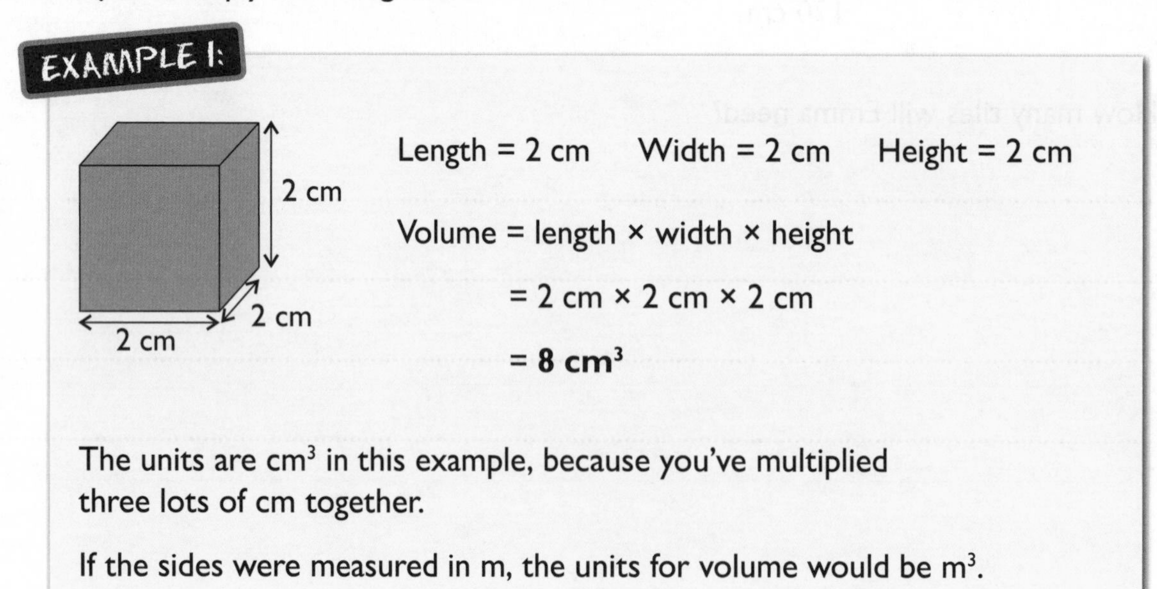

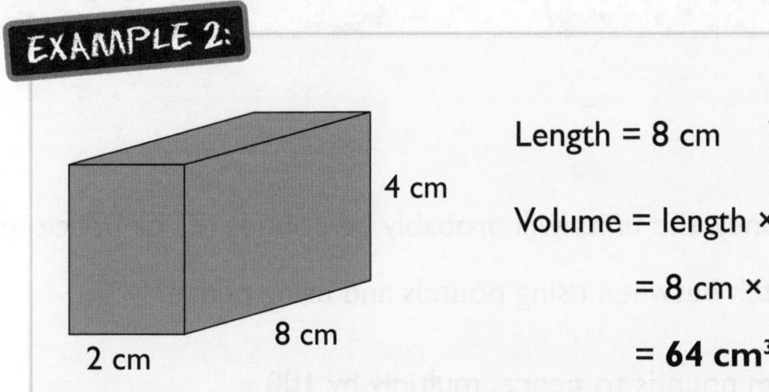

Length = 8 cm Width = 2 cm Height = 4 cm

Volume = length × width × height

= 8 cm × 2 cm × 4 cm

= **64 cm^3**

Practice Questions

1) What are the volumes of the shapes below?

a)

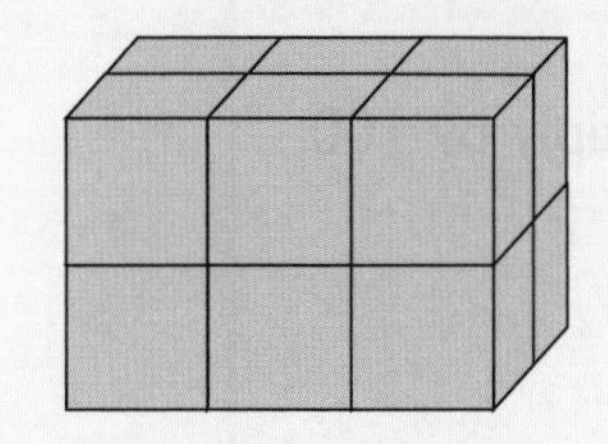

= 1 cm^3

....

b)

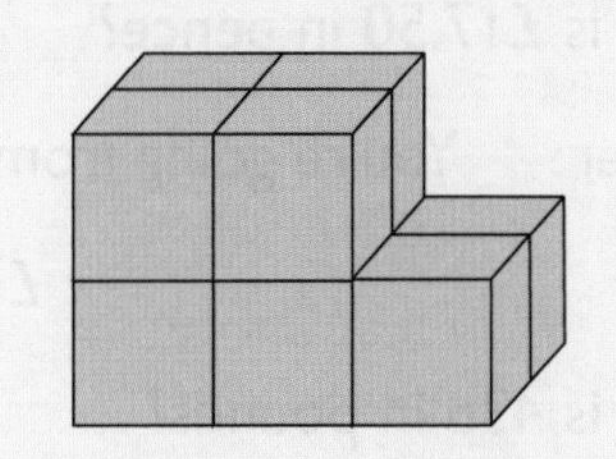

= 1 cm^3

....

2) Calculate the volumes of the shapes below.

a)

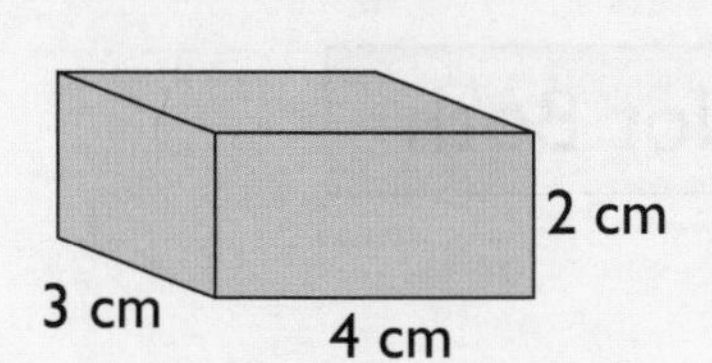

....

....

b)

4 cm

5 cm

2.5 cm

....

....

3) Sarah has bought a sand pit for her daughter. The sand pit is 0.8 m long, 0.8 m wide and 0.25 m high. What volume of sand can it hold? Give your answer in m^3.

....

....

Money

Pounds and Pence

1) If you get a question on money, the units will probably be pounds (£) or pence (p).
2) You need to be able to switch between using pounds and using pence.

To go from pounds to pence, multiply by 100.

To go from pence to pounds, divide by 100.

EXAMPLES:

What is £17.50 in pence?

Answer: You're going from pounds to pence, so multiply by 100.

£17.50 × 100 = **1750p**

What is 43p in pounds?

Answer: You're going from pence to pounds, so divide by 100.

43p ÷ 100 = **£0.43**

Use Pounds OR Pence in Calculations — Not Both

1) You may get a question that uses pounds and pence.
2) If you do, you'll need to change the units so that they're all in pounds or all in pence.

EXAMPLE:

Chris buys a sandwich for £3.49, a cup of tea for £1.25 and a packet of crisps for 78p. How much does he need to pay in total?

1) Change the price of the crisps from pence to pounds.

78p ÷ 100 = £0.78

2) All the prices are in the same units now (£), so just add them up.

£0.78 + £3.49 + £1.25 = **£5.52**

3) If the question tells you what units to give your answer in then make sure you use those. If it doesn't, you can change everything into pounds or into pence.

Value for Money Calculations

1) If you're buying a pack of something, you can work out how much you're paying for each item.

Price per item = total price ÷ number of items

2) You can then compare the price per item for that pack with other packs.

EXAMPLE:

A shop sells tins of beans in packs of 4 or 6. The 4-pack costs £1.80. The 6-pack costs £2.40.

4-pack: Price per tin = £1.80 ÷ 4 = £0.45

6-pack: Price per tin = £2.40 ÷ 6 = £0.40

Price per tin = total price ÷ number of tins

The **6-pack costs less per tin**, so it's **better value** than the 4-pack.

3) You can also compare costs by looking at how much you'd pay per gram of something.

Price per gram = total price ÷ number of grams

EXAMPLE:

A 400 g tin of beans costs 72p. A 150 g tin of beans costs 39p.

400 g tin: Price per gram = 72p ÷ 400 = 0.18p

150 g tin: Price per gram = 39p ÷ 150 = 0.26p

The **400 g tin costs less per gram**, so it's **better value** than the 150 g tin.

Practice Questions

1) a) What is £1.27 in pence? b) What is 219p in pounds?

....

2) A multipack of crisps costs £1.20. It has 6 packets in it. Another multipack costs £2.30 and has 10 packets in it. Which multipack is the best value for money?

....

....

3) A 50 g tin of caviar costs £63.00. A 30 g tin costs £39.00. Which is the best value?

....

....

Time

Time Has Lots of Different Units

You need to be able to use lots of different units for time. You also need to be able to change between them. Here are how some of the units of time are related:

60 seconds = 1 minute	7 days = 1 week	10 years = 1 decade
60 minutes = 1 hour	365 days = 1 year	100 years = 1 century
24 hours = 1 day	12 months = 1 year	

15 minutes = a quarter of an hour

30 minutes = half an hour

45 minutes = three quarters of an hour

The 12-Hour Clock and the 24-Hour Clock

1) You can give the time using the 12-hour clock or the 24-hour clock.
2) The 24-hour clock goes from 00:00 (midnight) to 23:59 (one minute before the next midnight).

EXAMPLE:

01:00 is 1 o'clock in the morning. 13:00 is 1 o'clock in the afternoon.

3) The 12-hour clock goes from 12:00 am (midnight) to 11:59 am (one minute before noon), and then from 12:00 pm (noon) till 11:59 pm (one minute before midnight).

EXAMPLE:

3 am is 3 o'clock in the morning. 3 pm is 3 o'clock in the afternoon.

4) For times in the afternoon, you need to add 12 hours to go from the 12-hour clock to the 24-hour clock. Take away 12 hours to go from the 24-hour clock to the 12-hour clock.

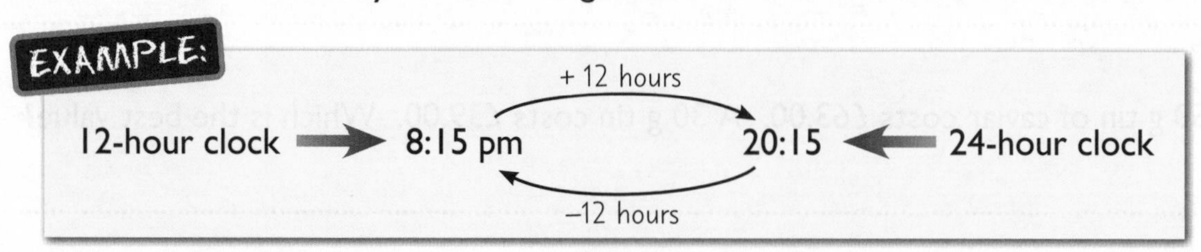

Practice Questions

1) How many minutes are there in 2 hours?

....

2) How many years are there in 3 decades?

....

3) How many minutes is 240 seconds?

....

4) Change the times below from the 24-hour clock to the 12-hour clock.

a) 09:00 b) 16:45

....

5) Change the times below from the 12-hour clock to the 24-hour clock.

a) 5:15 pm b) 7:05 am

....

6) Chloe gets to a bus stop at 10:47 pm. The last bus leaves at 22:55. Has she missed it?

....

Working Out Lengths of Times

To work out how long something took, break it into stages.

EXAMPLE:

Claire caught a train at 8:30 am and arrived at 11:17 am.
How long was her journey?

8:30 am → (30 mins) → 9:00 am → (2 hours) → 11:00 am → (17 mins) → 11:17 am

Add up the hours and minutes separately: 2 hours
30 mins + 17 mins = 47 mins

So the journey took **2 hours and 47 mins**.

Working Out Times

1) You may need to work out what time something will happen.
   For example, what time something will start or finish, or when to meet someone.
2) The best way to do this is to split the time into chunks.

EXAMPLE 1:

Melissa is visiting her sister. The drive takes 3 hours and 30 minutes.
She'll stop at a service station for a 20 minute break.
If she leaves at 6:00 pm, what time should she arrive?

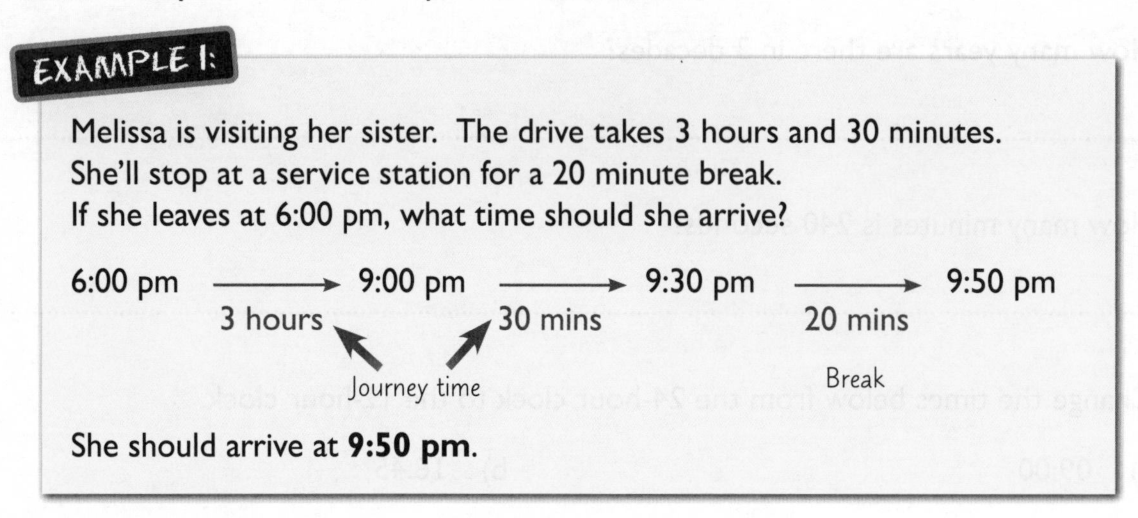

She should arrive at **9:50 pm**.

EXAMPLE 2:

Pranav is meeting a friend for lunch.

He has a half hour meeting with his bank manager at 11 am.

He has some shopping to do afterwards, which will take about 40 minutes.

It will then take him about 25 minutes to drive to the cafe for lunch.

What is the earliest time Pranav should arrange to meet his friend?

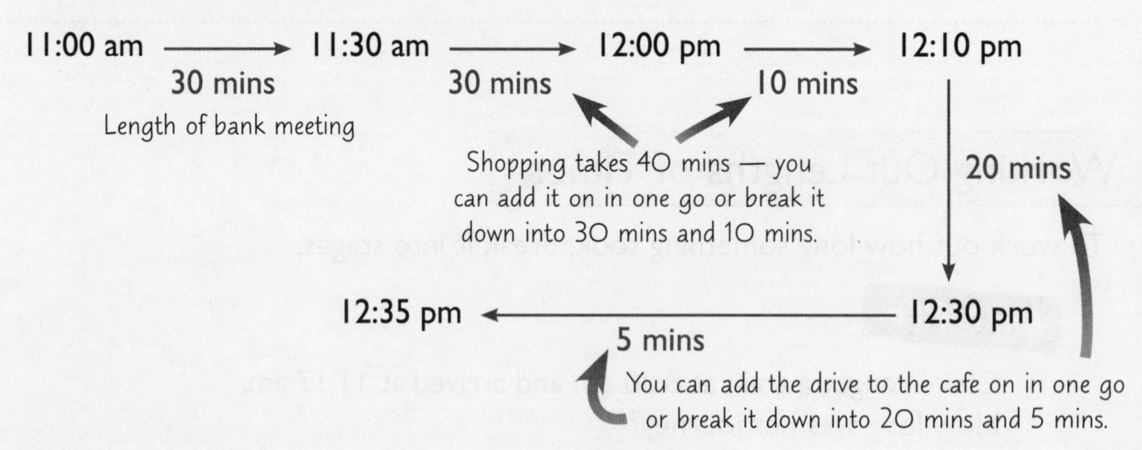

So the earliest he can be at the cafe is **12:35 pm**.
To be on the safe side he might arrange to meet at 12:45 pm.

Practice Questions

1) A film starts at 7:15 pm and finishes at 9:30 pm. How long is the film?

....

....

2) Dai catches a bus at 8:55 am and gets off at 11:47 am. How long was the bus trip?

....

....

3) Chris is cooking dinner. His recipe says it will take 35 minutes to prepare and 45 minutes to cook. If he starts at 6:30 pm, what time will dinner be ready?

....

....

4) Mary arrives at a museum at 3:35 pm. The museum closes at 5:30 pm.
How long will Mary be able to spend in the museum?

....

....

5) Flavia has 6 wedding invitations to make. It takes her 20 minutes to make each invitation.
She'll stop for half an hour to eat dinner. If she starts at 17:45, what time will she finish?

....

....

....

6) Ashley would like to go to the theatre to see a show that starts at 19:30.
She usually gets home from work at 5:45 pm.
She reckons it will take her half an hour to get ready, and 45 minutes to drive to the theatre.
She wants to leave herself a quarter of an hour to find a parking space.
Will she be able to get to the theatre in time?

....

....

....

Timetables

Timetables Have Information About When Things Happen

1) Timetables have columns and rows.
2) Columns are the strips that go up and down. Rows are the strips that go across.
3) There are lots of different types of timetables — the best way to learn how to use them is to practise.

EXAMPLE 1:

The timetable below shows train times.
What time would you need to leave Millom to get to Barrow for 15:00?

Millom	12:57	14:28	14:59	16:23
Green Road	13:02	14:35	15:03	16:27
Foxfield	13:05	14:38	15:07	16:31
Kirkby	13:09	14:43	15:11	16:35
Askam	13:14	14:51	15:16	16:40
Barrow	13:27	14:59	15:29	16:53

1) Find Barrow in the timetable.
2) Follow that row until you reach the last time before 15:00. It's 14:59.
3) Go up the column till you reach the top row — the leaving time from Millom.
4) So you'd need to leave Millom at **14:28**.

EXAMPLE 2:

Ayla has started a new job. The timetable for her first three days is below.

	Monday	Tuesday	Wednesday
9:00 am	Site tour	Training	Q and A session
11:00 am	Training	IT introduction	Training
12:30 pm	Lunch	Skills tests	Lunch
1:30 pm	Introduction to HR	Lunch	First shift
2:30 pm	Training	Training	Feedback

1) What time is lunch on Tuesday? Answer: **1:30 pm**
2) What will Ayla be doing at 14:30 on Monday? Answer: **Training**.
3) When is Ayla's first shift? Answer: **1:30 pm on Wednesday**.

You Need to be Able to Create Timetables

There are no set rules for making timetables. You just need to use the information that you're given and fit it together the best way you can.

Jess is at a film festival. She wants to see 3 films. The film times are below.

The Twitcher (90 mins): 10:20, 11:40, 13:45, 15:30

Fireball (120 mins): 10:15, 12:00, 14:00, 15:45

Cloudy Days (65 mins): 10:00, 13:00, 17:00

Jess needs to leave the festival by 17:30.
Draw a timetable for her day. Include time for lunch.

Answer:

There isn't just one right answer for this question. There are lots of different timetables that would work. Just make sure that...

- You include all three films.
- The films don't overlap.
- You put in a lunch break at a sensible time.
- The last film finishes before 17:30.

Here's an example of a timetable that would work:

	Start	Finish
Fireball	10:15	12:15
Lunch	12:15	13:00
Cloudy Days	13:00	14:05
The Twitcher	15:30	17:00

You know how long each film lasts, so you can work out what time they'll finish.

Make sure you put lunch at a sensible time, and that it's a long enough gap.

You may need to try a few different orders before you find one that works.

The last film needs to have finished by 17:30.

Practice Questions

1) Look at the timetable on the right.

Bus Timetable				
Bus Station	1845	1900	1915	1930
Market Street	1852	1907	1922	1937
Long Lane Shops	1901	1916	1931	1946
Train Station	1911	1926	1941	1956
Hospital	1923	1938	1953	2008

a) What time would you need to get the bus from the bus station to get to the train station for 19:30?

....

b) What time would you need to catch the bus from the bus station to get to Market Street for 7:00 pm?

....

2) Mark is a hairdresser. His bookings for the week are shown below. Tia wants to book a cut and colour. A cut and colour will take 90 minutes. Suggest a time that Mark could fit Tia in.

	Mon	Tue	Wed	Thur	Fri
9:00	Sally (cut)		Mary (cut)	Milly (cut)	Rhian (wedding)
10:00	Jon (cut)	Jane (cut and colour)	Delene (cut)	Atiya (cut)	
11:00				Rhian (trial up-do)	
12:00	Lunch		Freya (colour)	Lunch	
13:00		Lunch			Lunch
14:00	Ali (cut)		Lunch		Sami (cut)
15:00				Emma (cut)	

....

3) The speakers for a two day conference are listed below. The lengths of their talks are given in brackets. Plan a timetable for the conference. Both days will start at 10:00 am and finish by 3:00 pm. Leave a 1 hour lunch break and at least 10 minutes between each talk.

K Craig (1 hour 15 mins)	M Falkner (2 hours)	H Gregson (25 mins)	M Tyler (50 mins)
M Hamill (1 hour 30 mins)	S Williams (30 mins)	J Towle (45 mins)	

Angles

Angles Measure How Far Something Has Turned

1) Angles tell you how far something has turned from a fixed point. The bigger the angle, the bigger the turn.
2) Angles are measured in degrees (°).

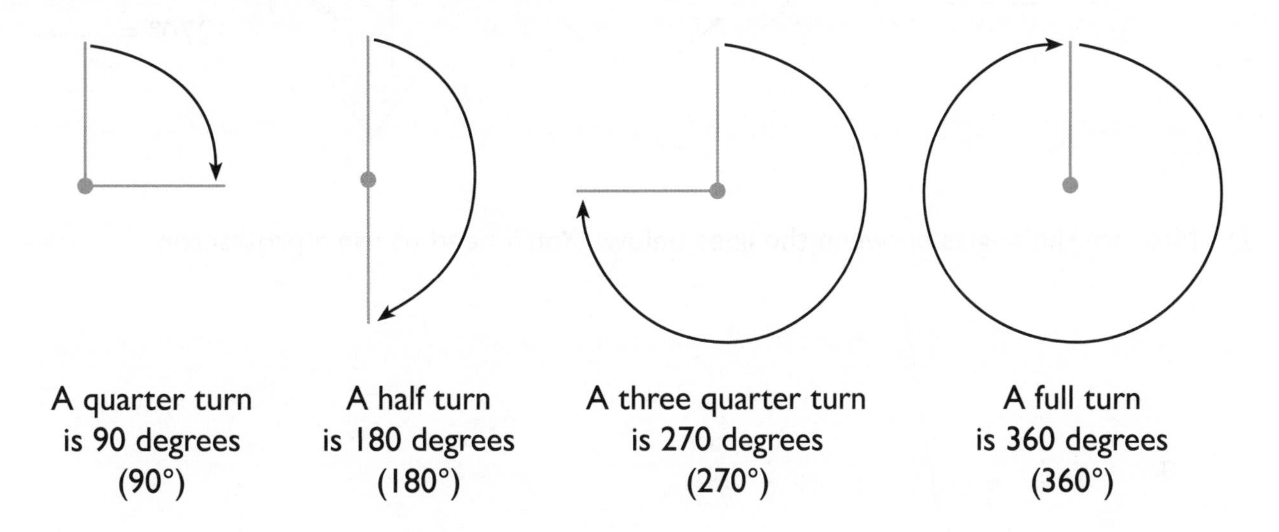

A quarter turn is 90 degrees (90°)

A half turn is 180 degrees (180°)

A three quarter turn is 270 degrees (270°)

A full turn is 360 degrees (360°)

You Can Measure Angles Between Lines

You can use a protractor to measure angles of up to 180°.

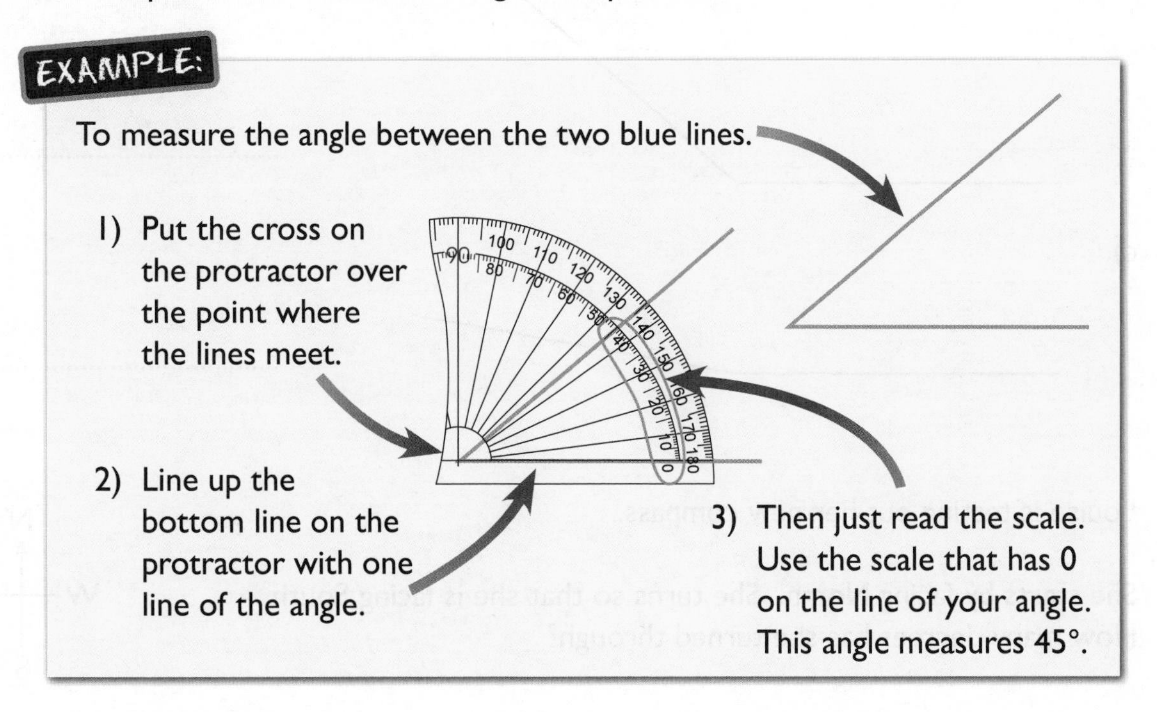

Practice Questions

1) In the diagrams below three arrows have been turned.
Write the letter of the arrow next to the number of degrees it has turned.

A B C

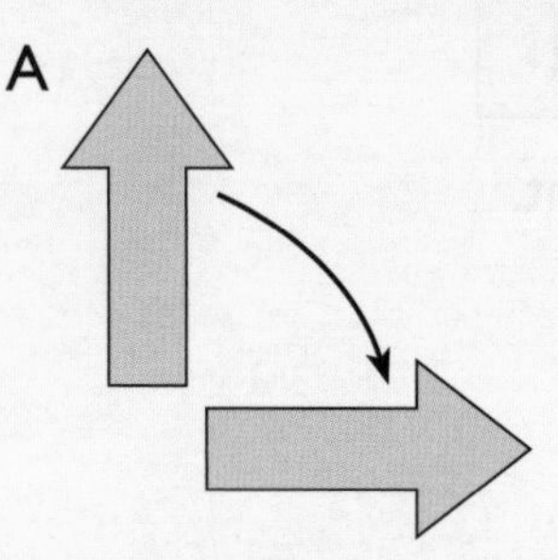

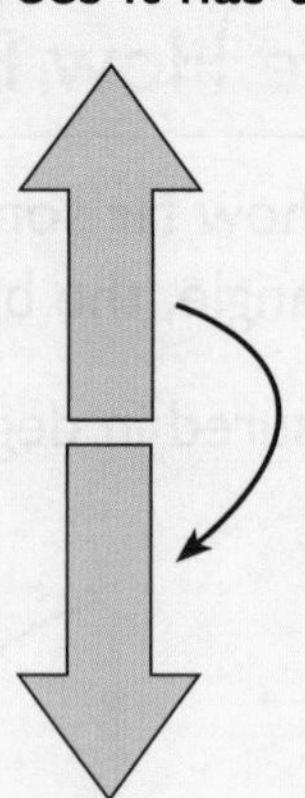

90° = ...

180° = ...

270° = ...

2) Measure the angles between the lines below. You'll need to use a protractor.

a)

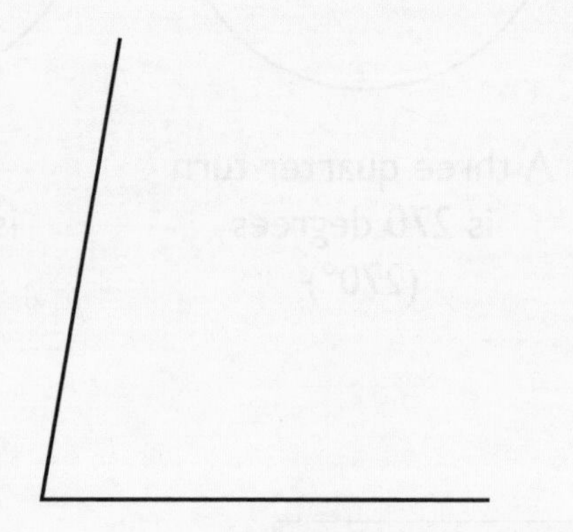

....

b)

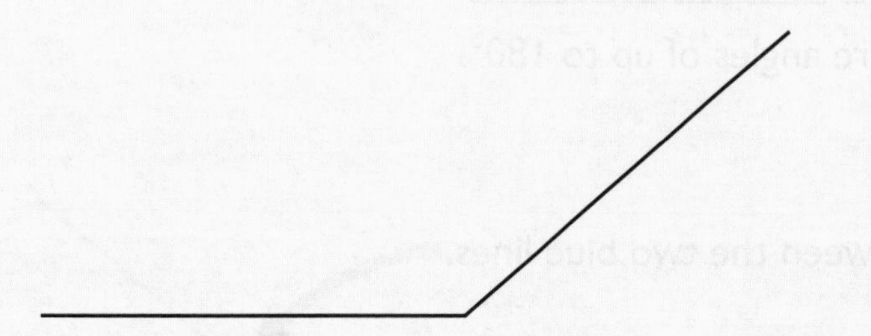

....

c)

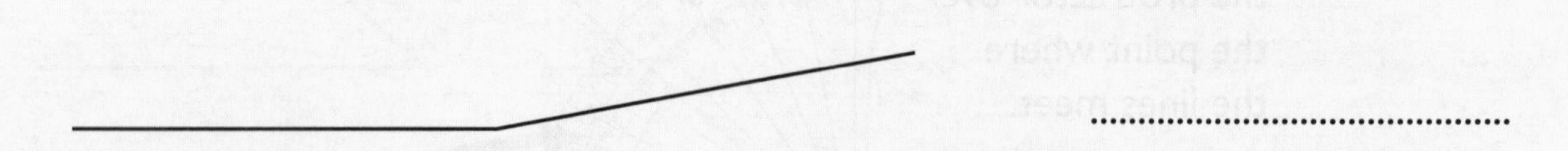

....

3) Louise is testing out her new compass.

She starts by facing North. She turns so that she is facing South.
How many degrees has she turned through?

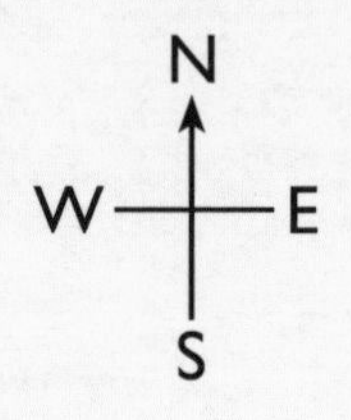

....

Symmetry and Tessellation

Some Shapes Have Lines of Symmetry

1) Shapes with a line of symmetry have two halves that are mirror images of each other.
2) You could fold a shape along this line and the sides would fold exactly together.
3) Some shapes have more than one line of symmetry.

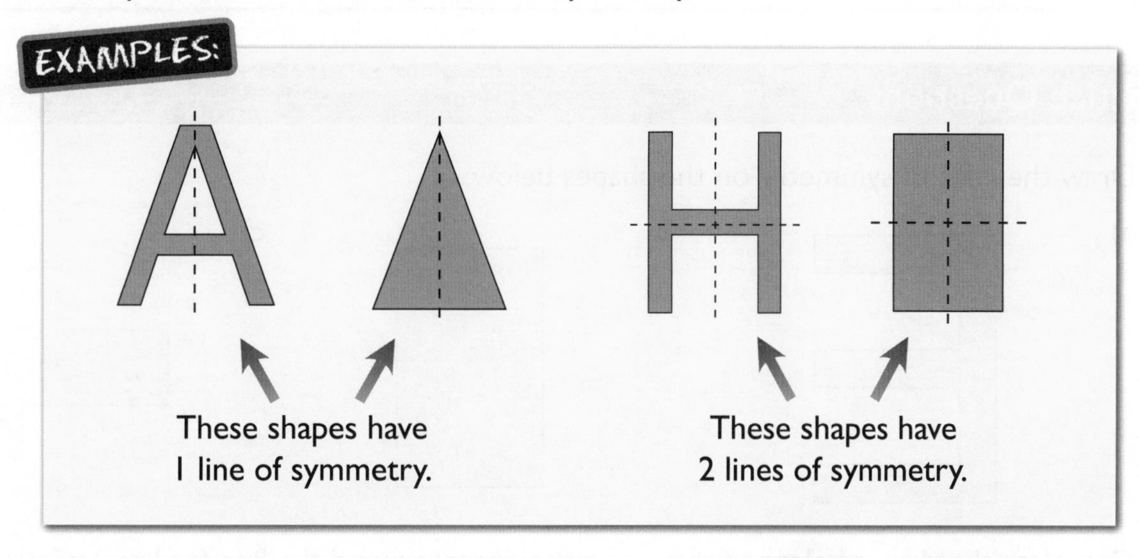

4) Some shapes have no lines of symmetry.

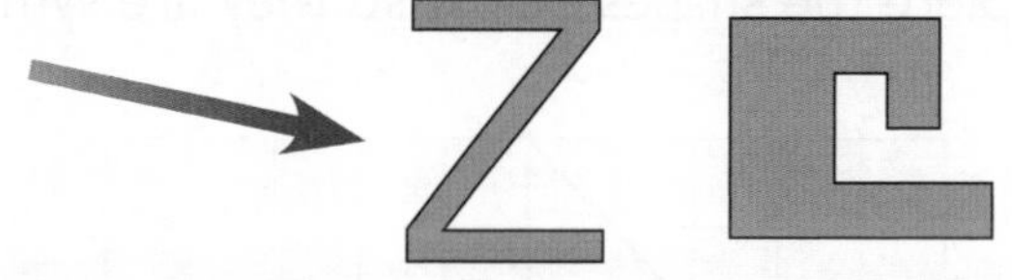

When Shapes Fit Together We Say They Tessellate

1) Tessellation is where shapes are put together in a pattern to leave no gaps.
2) Floor tiles tessellate — they fit together without gaps or overlaps.

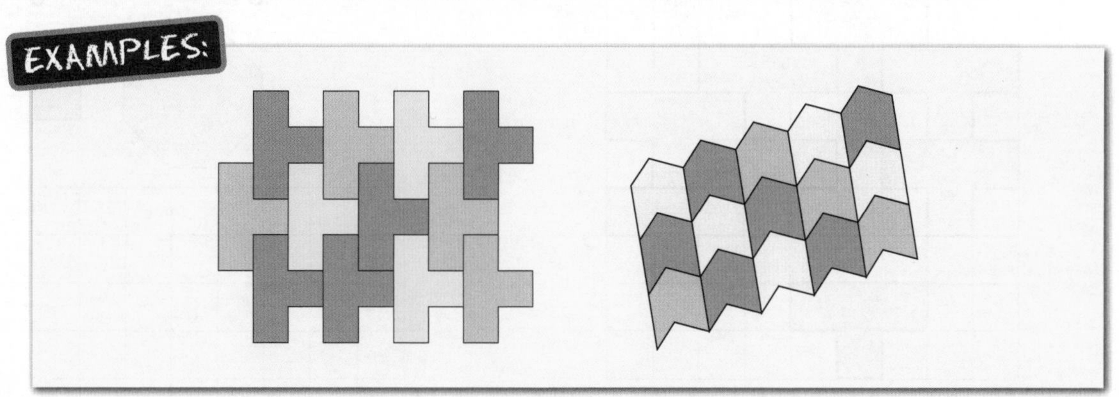

Different Shapes can Tessellate

Different shapes can be put together to make a tessellating pattern.

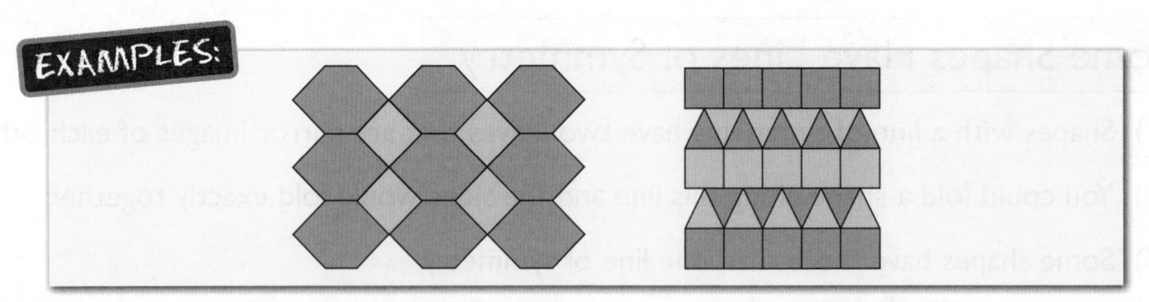

Practice Questions

1) Draw the lines of symmetry on the shapes below:

a)

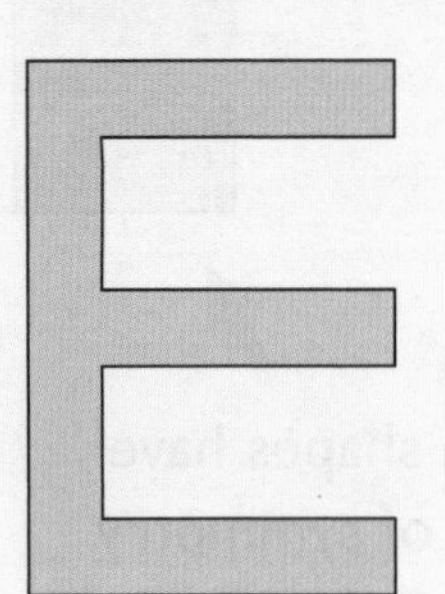

b)

c)

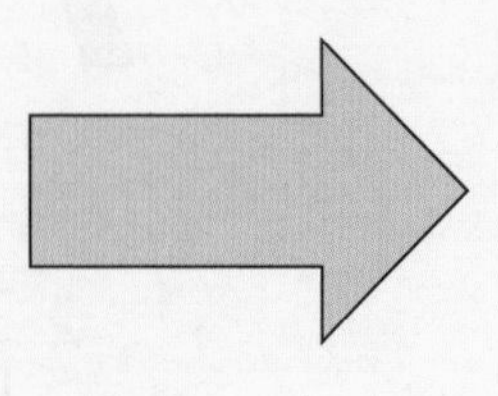

2) Complete the shapes below so they are symmetrical around the line (or lines) of symmetry.

a)

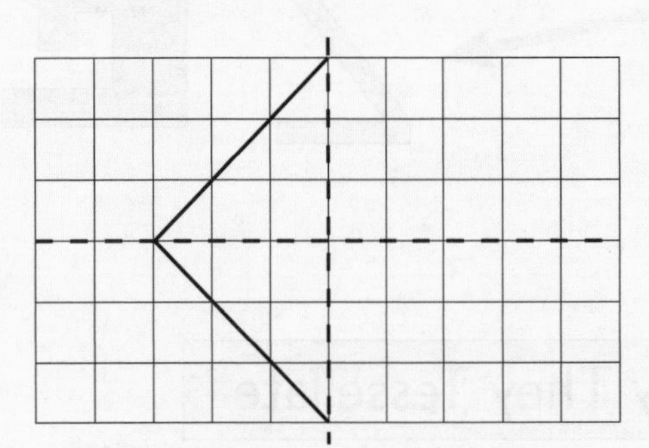

b)

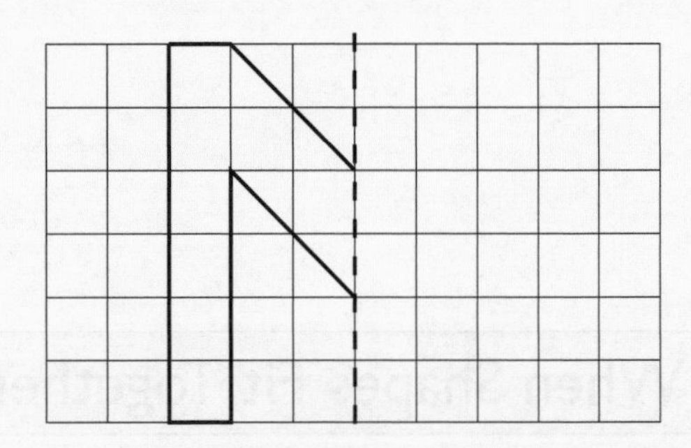

3) a) Complete the pattern in the grid below.

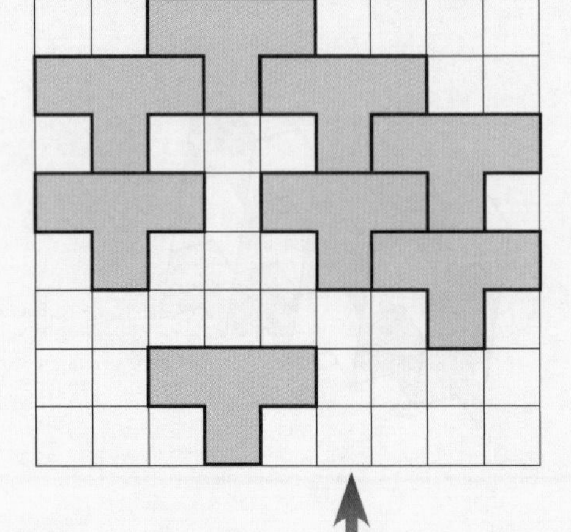

There will be some empty squares at the edges once you've finished.

b) Using both the shapes below, draw a tessellating pattern in the grid.

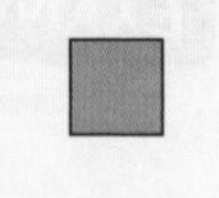

Plans

Plans Show How Things are Laid Out in an Area

1) A plan shows the layout of an area. For example, a plan might show a room and all the objects in it.

2) Plans are drawn as if you are looking down on the area from above — a bird's eye view.

A plan of a garden is shown below:

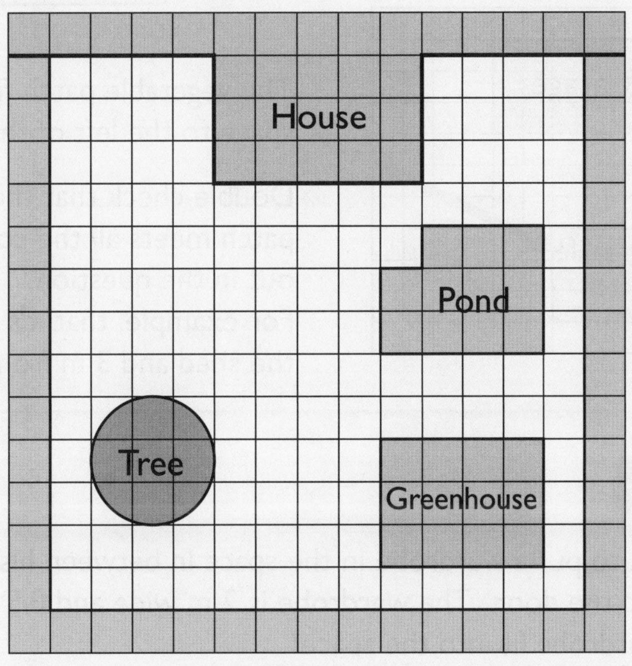

On this plan the length (and width) of each square on the grid is 1 m.

From this you can work out how large objects are and how far away things are from each other.

The greenhouse is 4 squares long and 3 squares wide.
This means it is 4 m long and 3 m wide.

There is a gap of two squares between the pond and the greenhouse.
So the gap is 2 m.

Using Plans

Plans are useful for deciding where a new object will fit in an area.

EXAMPLE 1:

Add a vegetable patch to the garden plan on the right.
The patch must be 2 m from the shed, 3 m from the garage and 1 m from the edge of the garden.

The length (and width) of each square on the grid is equal to 1 m in the garden.

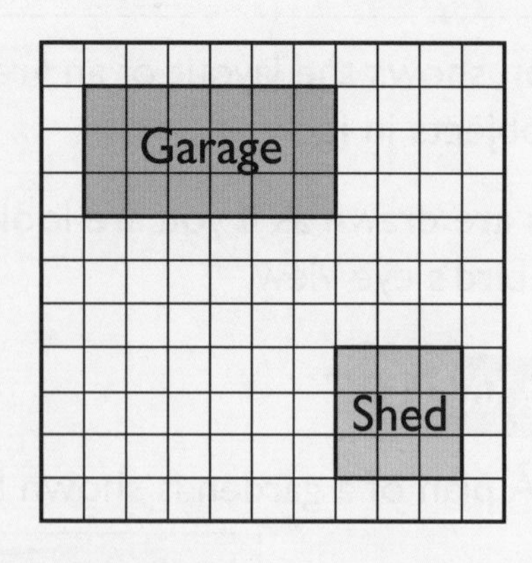

The vegetable patch fits into the space to the left of the shed.

Double check that the vegetable patch meets all the conditions set out in the question.
For example, that it's 2 m from the shed and 3 m from the garage.

EXAMPLE 2:

Gino wants to put a wardrobe in the space in between his bedroom window and the door. The wardrobe is 2 m wide and 0.5 m deep. Will the wardrobe fit into the space?

The length (and width) of each square on the grid is equal to 0.5 m in the bedroom.

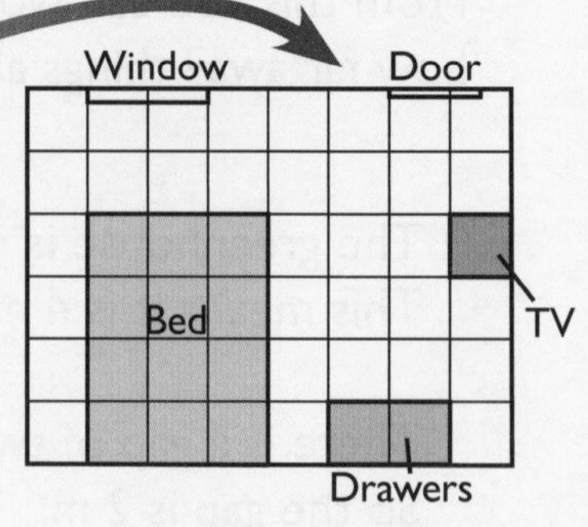

1) First, find the space Gino wants to put his wardrobe in.
2) Now work out how big the space is in metres:
   3 squares at 0.5 m each = 3 x 0.5 m = 1.5 m
   So the space is 1.5 m wide.
3) The wardrobe is 2 m wide, so it **won't** fit into this space.

Practice Questions

1) Marina wants to buy a new freezer. It's 1 m wide and 1 m deep.
She wants to put it between the bin and the end of the kitchen worktop.

The length (and width) of each square on the plan is equal to 0.5 m

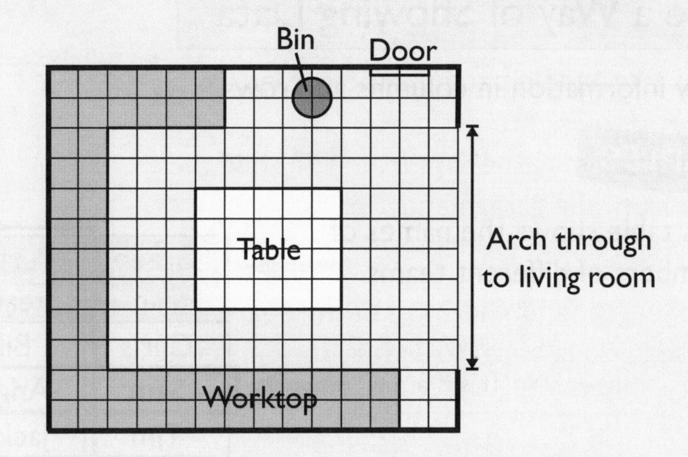

a) Will the freezer fit in the space?

....

b) Think of another place the freezer could go and draw it on the plan above.
The freezer must be against a wall.

2) The local council has bought a new climbing frame for a children's play area.

The climbing frame is 3 m long and 3 m wide.
It must be at least 3 m away from any other equipment.

The length (and width) of each square on the plan is equal to 1 m in the play area.

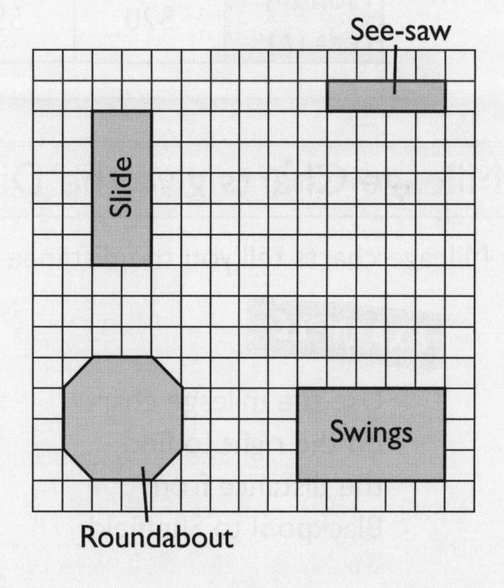

a) Draw a place where the climbing frame could be built on the plan above.

b) The council also has a larger climbing frame. It is 4 m long and 4 m wide.
Could this go in the place you chose in part a)? Explain your answer.

....

....

Tables

Tables are a Way of Showing Data

Tables show information in columns and rows.

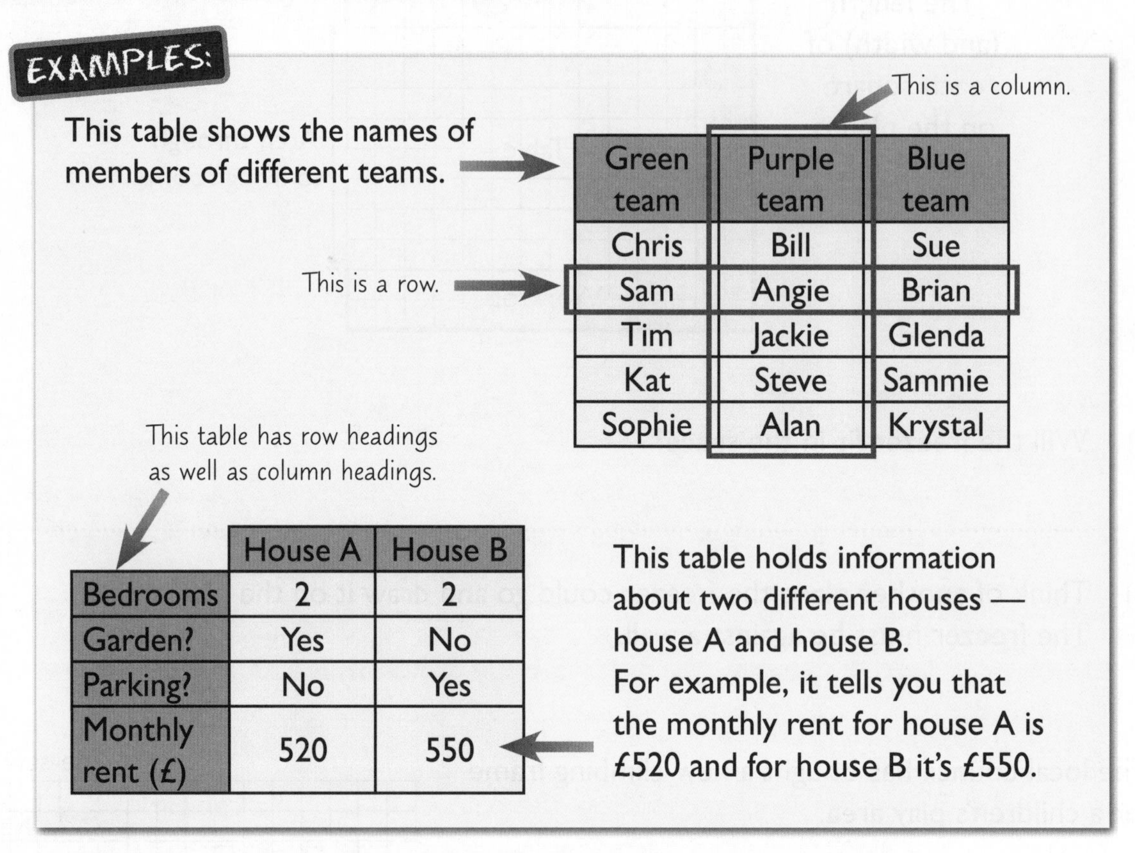

Green team	Purple team	Blue team
Chris	Bill	Sue
Sam	Angie	Brian
Tim	Jackie	Glenda
Kat	Steve	Sammie
Sophie	Alan	Krystal

	House A	House B
Bedrooms	2	2
Garden?	Yes	No
Parking?	No	Yes
Monthly rent (£)	520	550

Mileage Charts give the Distances Between Places

Mileage charts tell you the distance between different places.

Use the mileage chart on the right to find the distance from Blackpool to Sheffield.

Blackpool			
52	Manchester		
56	34	Liverpool	
92	40	72	Sheffield

Distances are shown in miles.

1) Move down the column underneath the 'Blackpool' heading and across the row next to the 'Sheffield' heading.
2) Where the column and the row meets is the distance between the two places — **92 miles**.

Blackpool			
52	Manchester		
56	34	Liverpool	
92	40	72	Sheffield

Below is another type of mileage chart you might see.

EXAMPLE:

Use the mileage chart below to find the distance from Blackpool to Sheffield.

1) Find 'Blackpool' on one side of the chart and 'Sheffield' on the other.
2) Move across the row and down the column to the square where the row and the column meet.
3) It doesn't matter if you've followed the path of the green arrows or the black ones — the answer will be **92 miles**.

	Blackpool	Manchester	Liverpool	Sheffield
Blackpool		52	56	92
Manchester	52		34	40
Liverpool	56	34		72
Sheffield	92	40	72	

Distances are shown in miles.

Drawing Tally Charts and Frequency Tables

You can use a tally chart to put data into different categories.

EXAMPLE:

You could use a tally chart to show the colours of cars in a car park:

There are 2 blue cars.
There are 3 red cars.
There is 1 green car.
There are 6 silver cars.
There are 3 white cars.

Colour	Tally
Blue	II
Red	III
Green	I
Silver	~~IIII~~ I
White	III

If another red car was seen in the car park you would add another line (tally mark) to the tally column next to red.

In a tally, every 5th mark crosses a group of 4 like this: ~~IIII~~
So ~~IIII~~ I represents 6 (a group of 5 plus 1).

You can add another column to make a frequency table.
You fill this in by adding up the tally marks for each colour.

Check the frequencies — the total should be the same as the number of tally marks (cars).

Colour	Tally	Frequency
Blue	II	2
Red	III	3
Green	I	1
Silver	~~IIII~~ I	6
White	III	3
		Total: 15

Making a Table for a Set of Data

Tables are useful for organising data so that it's easy to understand.

EXAMPLE:

10 friends are voting for which takeaway they should order from a choice of Chinese, Indian, Mexican and Thai.

Here are the results:

Nick — Thai, Sarah — Indian, Rikki — Indian, Kate — Thai, Arnold — Indian, Mark — Thai, Tom — Chinese, Mike — Thai, Helen — Mexican, Matt — Thai.

These results can be put into a table.

The table will need space for all the takeaway options, each person's vote and the totals for each option.

Here's an example of a table to show these results:

Takeaway option	Tally	Frequency
Chinese	I	1
Indian	III	3
Mexican	I	1
Thai	~~IIII~~	5
		Total: 10

1) Each tally mark shows one person's vote.
2) You can check all the results have been tallied by looking at the total frequency — it should match the number of people asked (10).

The type of table you need depends on the information you need to display.

Practice Questions

1) The table below contains facts about three different cars. Use the table to answer the following questions.

	Car A	Car B	Car C
Engine size (litres)	1.6	1.2	1.4
0-60 MPH (seconds)	9.7	14.1	12.2
Fuel consumption (mpg)	35	58	43
Price (£)	9867	6492	12 087

a) What is the engine size of car C?

....

b) What is the 0-60 MPH time of car B?

....

c) What is the price of car A?

....

Practice Questions

1) Use the mileage chart below to answer the following questions.
Distances are shown in miles.

London			
65	Bognor Regis		
292	354	Millom	
413	456	166	Edinburgh

a) What is the distance between Millom and London?

....

b) What is the distance between Bognor Regis and Edinburgh?

....

c) Jinden is driving from Edinburgh to Millom and then driving to London.
What is the total mileage for his journey?

....

2) A groundsman is collecting information about the types of tree in a park. He walks past a pine tree, a yew, another yew, two cedars, an oak, two beech, another oak, a yew, a cedar, a beech and finally two more cedars.

a) Complete the table on the right using the information above.

Type of tree	Tally	Frequency
Beech		
Oak		
Yew		
Cedar		
Pine		
	Total	

b) How many cedar trees are there?

....

c) How many oak trees are there?

....

3) A cafe owner wants to keep track of how many drinks were sold in a weekend.
On Saturday 75 cups of tea, 60 cups of coffee and 86 fizzy drinks were sold.
On Sunday 60 cups of tea, 41 cups of coffee and 59 fizzy drinks were sold.
Draw a table to show this information. Include the total number of drinks sold on each day.

Charts and Graphs

Bar Charts Let You Compare Data Easily

1) A bar chart is a simple way of showing information.
2) On a bar chart you plot your data using two lines called axes (if you're talking about just one then it's called an axis).

EXAMPLE:

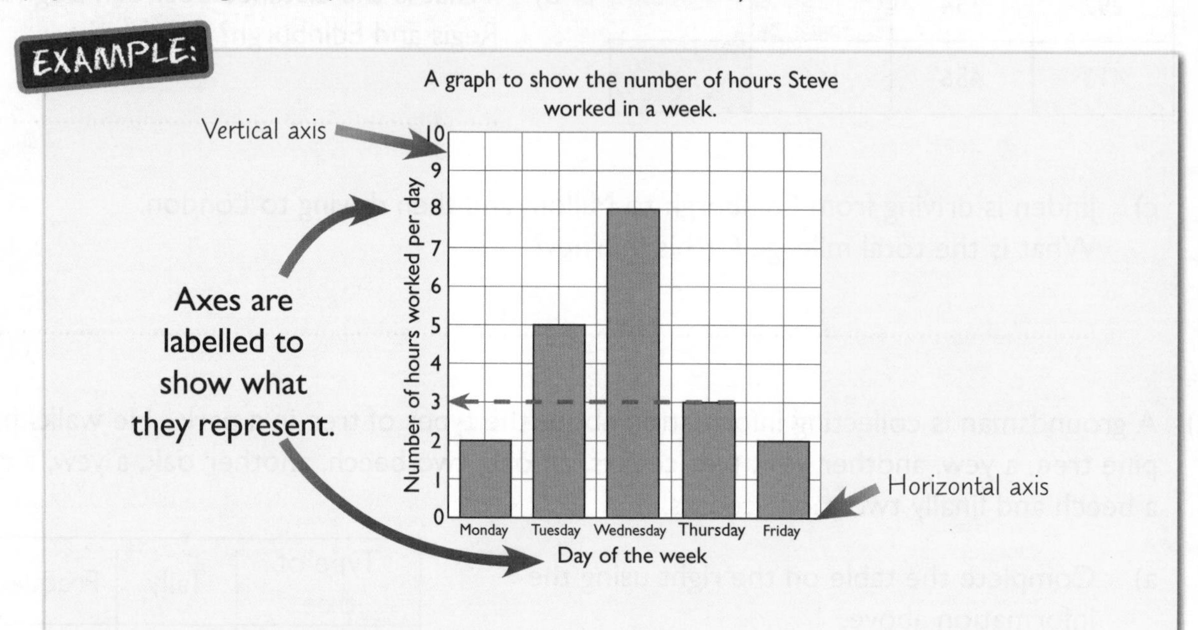

1) The height of each bar shows how many hours were worked each day.
2) Just read across from the top of the bar to the number on the vertical axis. For example, on Thursday Steve worked three hours.
3) You can draw conclusions from the chart. For example, Steve worked the most hours on Wednesday as it's the day with the tallest bar.

Practice Questions

1) The bar chart shows the number of different colours of shirt that were sold in a shop.

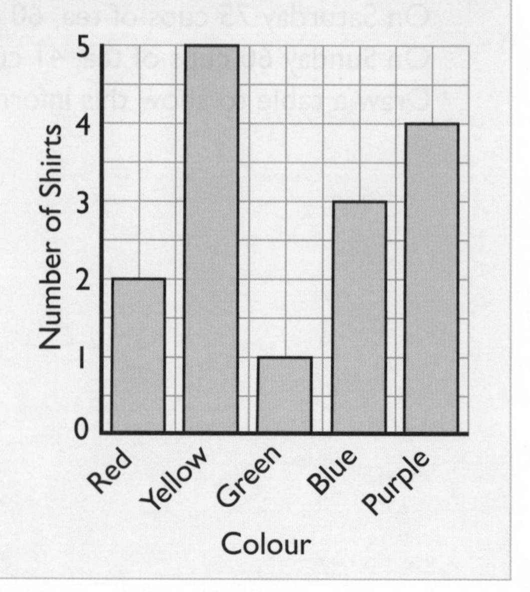

a) How many blue shirts were sold?

b) Which colour was the most popular?

c) How many shirts were sold altogether?

Line Graphs Show the Relationship Between Two Things

Line graphs are similar to bar charts but instead of bars, a line is used to show the data.

EXAMPLE 1:

The line graph below shows how the value of a car changes over time.

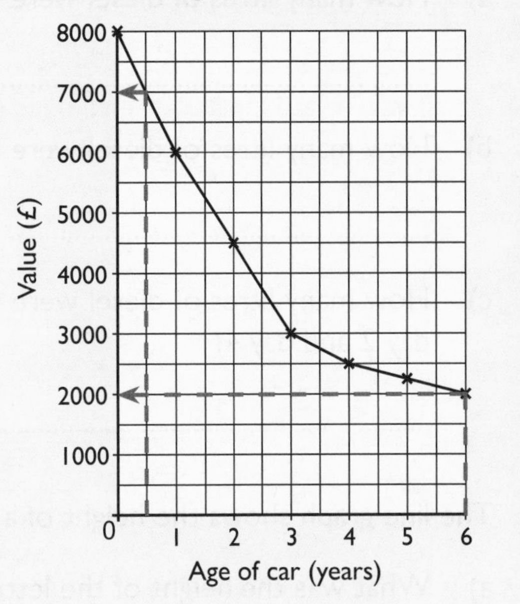

1) Line graphs are useful for showing how things change over time. For example, the graph on the right shows that the value of a car goes down as it gets older.
2) Each cross represents a data point. For example, the cross at the lowest point on the line shows that a 6-year old car is worth £2000.
3) You can use the graph to find out the value of the car at any time up to 6 years. For example, after 6 months the value is £7000.

EXAMPLE 2:

Here's a line graph for changing temperatures from degrees Celsius (°C) to degrees Fahrenheit (°F), or the other way round.

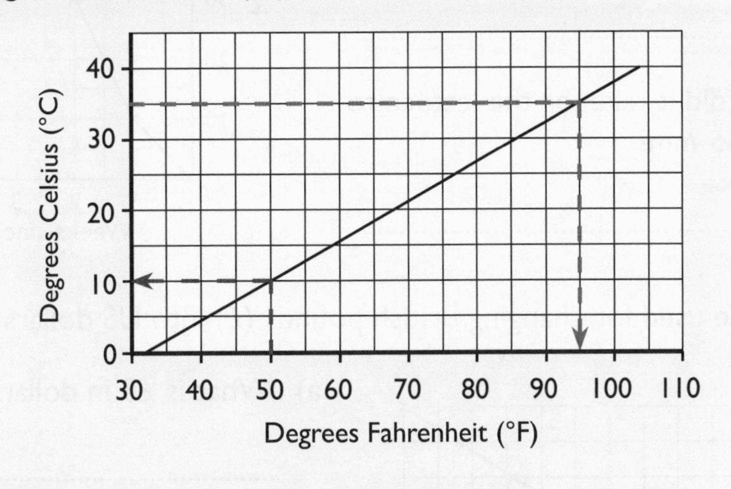

Sometimes line graphs don't have crosses — there's just a line to show the data.

What is 35 °C in degrees Fahrenheit (°F)?

1) From 35 °C on the vertical axis move across until you get to the line.
2) Go directly down to the horizontal axis.
3) The value on the horizontal axis is the answer — 95 °F.

You can also change a temperature from °F to °C — from the blue arrow you can see that 50 °F is the same as 10 °C.

Practice Questions

1) Miranda runs a village petrol station. This graph shows how the amount of diesel at the station decreased over 6 days.

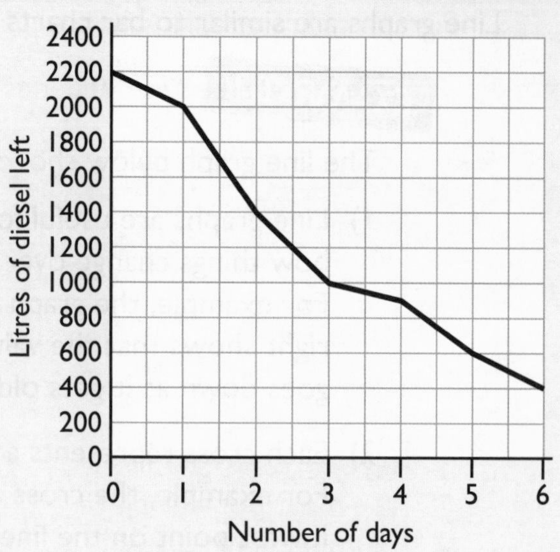

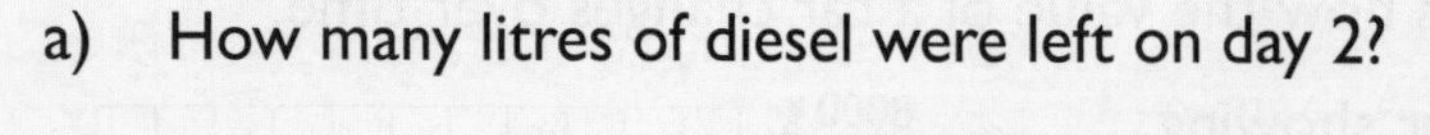

a) How many litres of diesel were left on day 2?

....

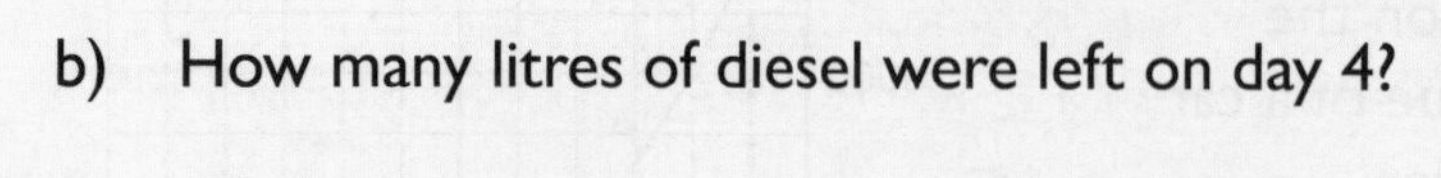

b) How many litres of diesel were left on day 4?

....

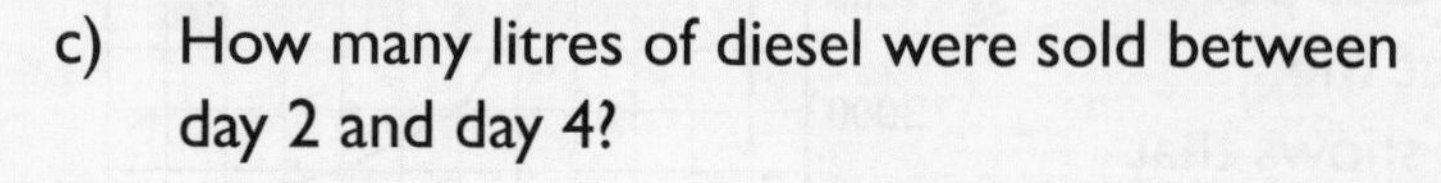

c) How many litres of diesel were sold between day 2 and day 4?

....

2) The line graph shows the height of a lettuce for 6 weeks after the seed was planted.

a) What was the height of the lettuce at week 3?

....

b) How many mm did the lettuce grow between week 3 and week 6?

....

c) How many weeks did it take for the lettuce to reach a height of 55 mm?

....

3) The graph below can be used for changing British pounds (£) into US dollars ($).

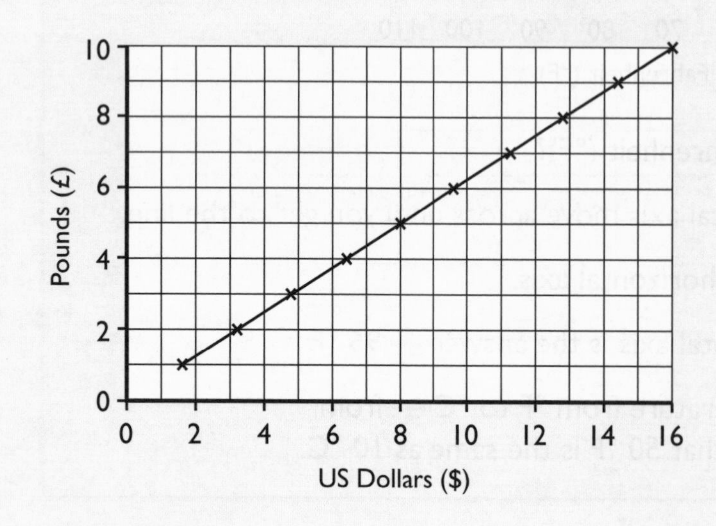

a) What is £5 in dollars?

....

b) What is $12 in pounds?

....

c) What is $13 to the nearest pound?

....

Pie Charts Show How Something is Split Up

1) Pie charts are circular and are divided into sections.
2) The size of each section depends on how much or how many of something it represents.

EXAMPLE:

This pie chart shows the most popular equipment at a gym.

The size of each section shows how many people prefer that piece of equipment.

This section is the biggest, so treadmills are the most popular piece of equipment.

It's ½ (50%) of the chart. This means that ½ of the people questioned prefer the treadmill.

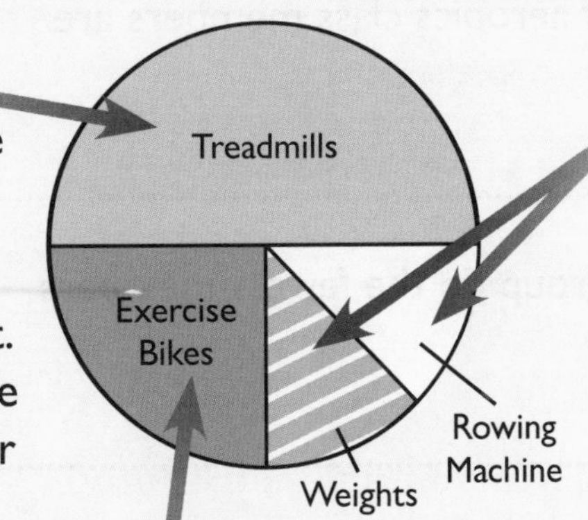

These are the smallest sections on the chart.

This means that the rowing machine and weights are less popular than both the treadmills and exercise bikes.

This section is ¼ (25%) of the chart. This means that ¼ of the people questioned prefer the exercise bikes.

Pictograms Use Pictures to Represent Numbers

1) Pictograms use pictures to show how many of something there are.
2) In a pictogram, each picture or symbol represents a certain number of items.

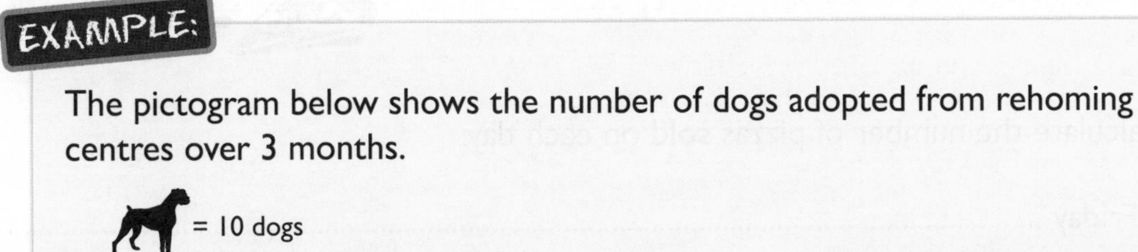

EXAMPLE:

The pictogram below shows the number of dogs adopted from rehoming centres over 3 months.

= 10 dogs

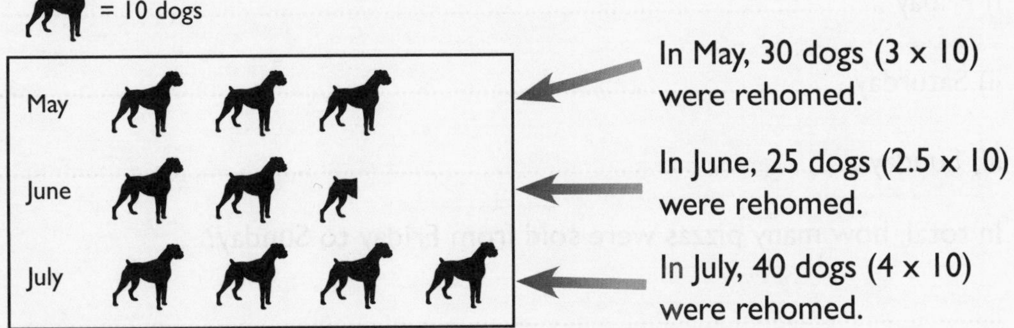

In May, 30 dogs (3 x 10) were rehomed.

In June, 25 dogs (2.5 x 10) were rehomed.

In July, 40 dogs (4 x 10) were rehomed.

Total dogs rehomed = 30 + 25 + 40 = **95 dogs**.

Practice Questions

1) The manager of a leisure centre is looking at the ages of members going to yoga and aerobics classes.

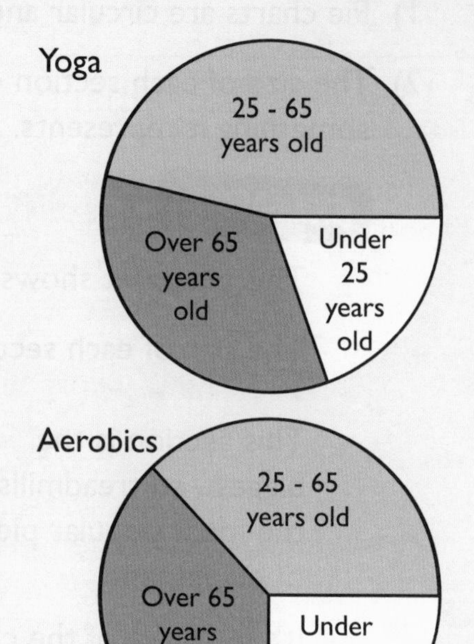

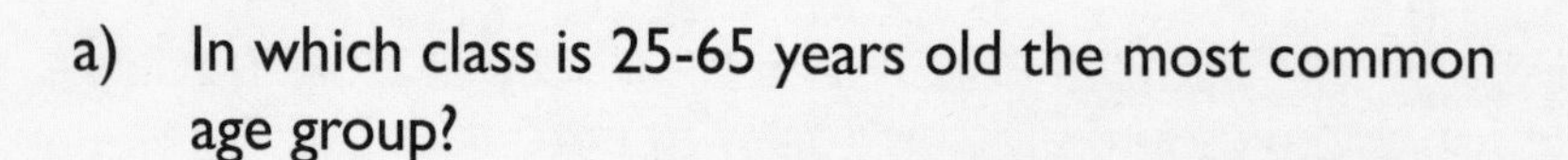

a) In which class is 25-65 years old the most common age group?

....

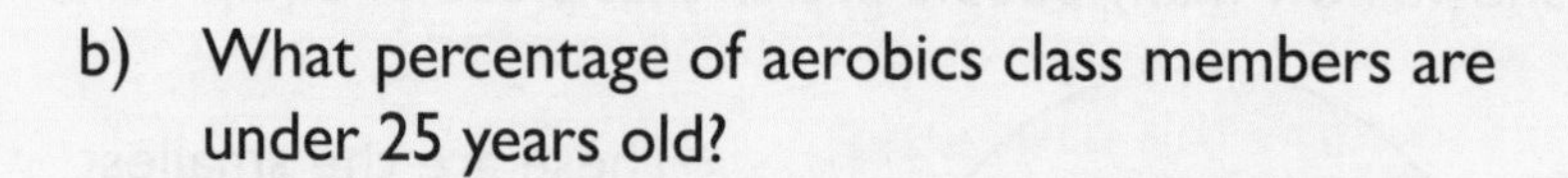

b) What percentage of aerobics class members are under 25 years old?

....

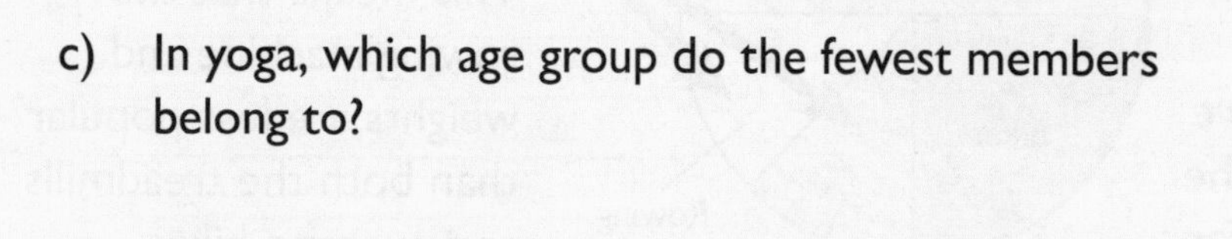

c) In yoga, which age group do the fewest members belong to?

....

2) The pictogram below shows the number of pizzas sold by Ronnie's Pizza over a weekend.

a) Calculate the number of pizzas sold on each day.

i) Friday ....

ii) Saturday ....

iii) Sunday ....

b) In total, how many pizzas were sold from Friday to Sunday?

....

Drawing Charts and Graphs

Drawing Bar Charts

You need to know how to draw a bar chart. The main steps are choosing what the axes will represent, choosing a scale for the axes and plotting (drawing) the data.

EXAMPLE:

The table below shows how many parking tickets a traffic warden gave out in a week. Draw a bar chart to show this data.

Day of the week	Number of tickets given out
Monday	22
Tuesday	18
Wednesday	8
Thursday	10
Friday	20

1) The bar chart will need to show the days of the week and the number of parking tickets given out. So these are what the axes will represent.

2) Work out a scale for the axes. (This is how the units will be spaced out along each axis.)

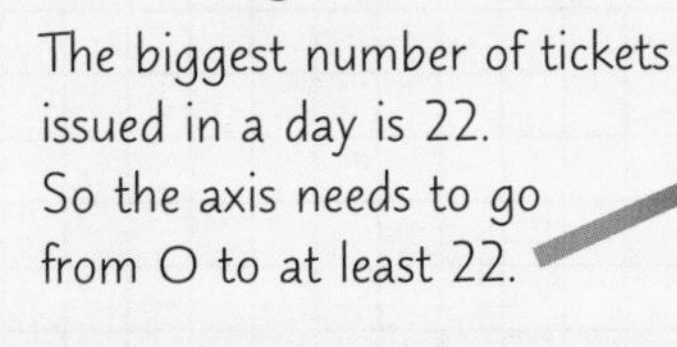

Choose how many tickets each square will represent. Here, 1 square = 2 tickets.

Add labels to your axes so that it's clear what they show.

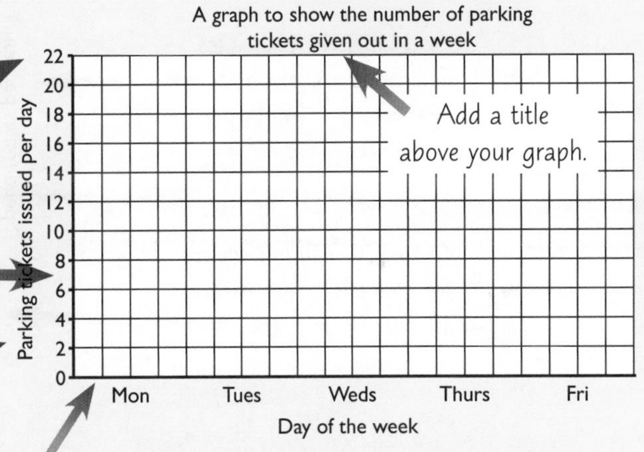

You need to pick a sensible scale for this axis. In this example, each bar will be 2 squares wide, with a 2 square gap between each bar.

3) Use a ruler to draw on the bars. Make sure that all the bars are the same width, and that the gaps between the bars are all the same size.

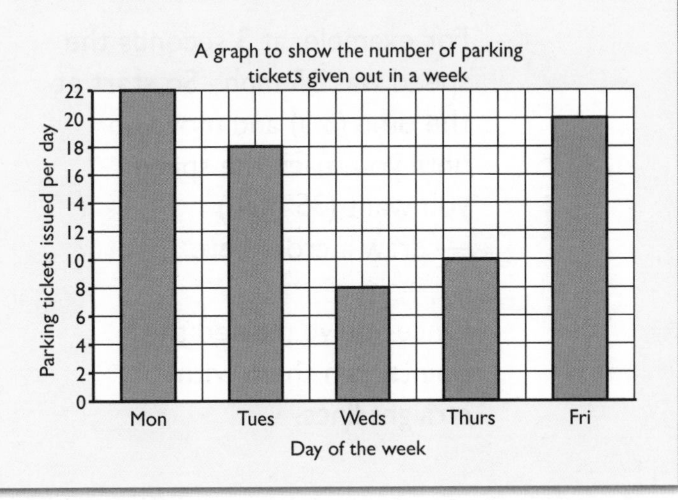

Drawing Line Graphs

The main steps for drawing line graphs are choosing what the axes will represent, choosing a scale for the axes and plotting the points.

EXAMPLE:

The speed of a car was measured every three seconds for 15 seconds. The speeds are shown in the table. Plot the speeds on a line graph.

1) The line graph will need to show the time and the speed of the car. So speed and time are what the two axes will represent.

Time (seconds)	Speed (miles per hour)
0	0
3	35
6	65
9	90
12	105
15	115

2) Work out the scales for the axes.

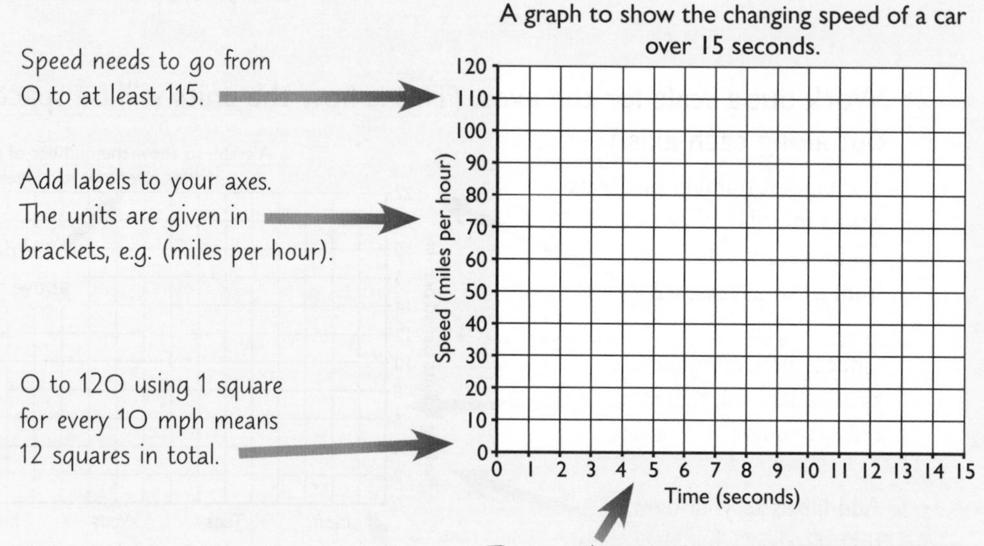

3) Now plot the points.

For example, at 3 seconds the speed was 35 mph. So start at the time (3 s) and move up until you reach the speed you want (35 mph) — draw a cross here.

Once you've plotted the points, join them with straight lines.

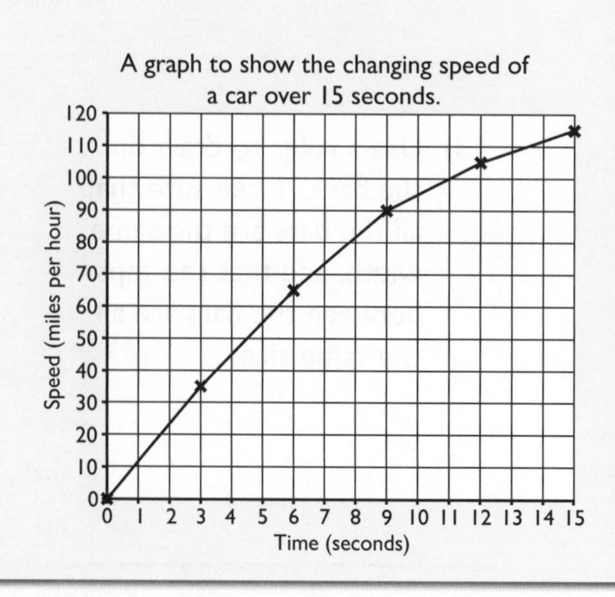

Practice Questions

1) A local council has built a new library. They asked local people how they felt about the library. The unfinished bar chart below shows some of the results.

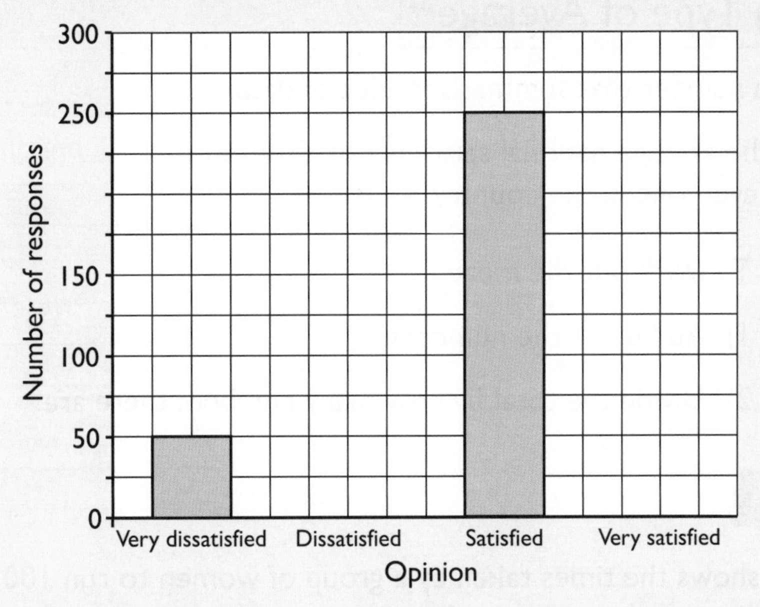

a) How many people said they were 'satisfied' with the new library?

....

b) 100 people who answered the survey said they were 'dissatisfied' by the new library. Add this bar to the chart.

c) 125 people who answered the survey said they were 'very satisfied' by the new library. Add this bar to the chart.

d) There is a number label missing from the vertical axis. Add this to the chart.

2) The average monthly rents in a town over 20 years are shown in the table below. Draw a line graph to show how rents have changed.

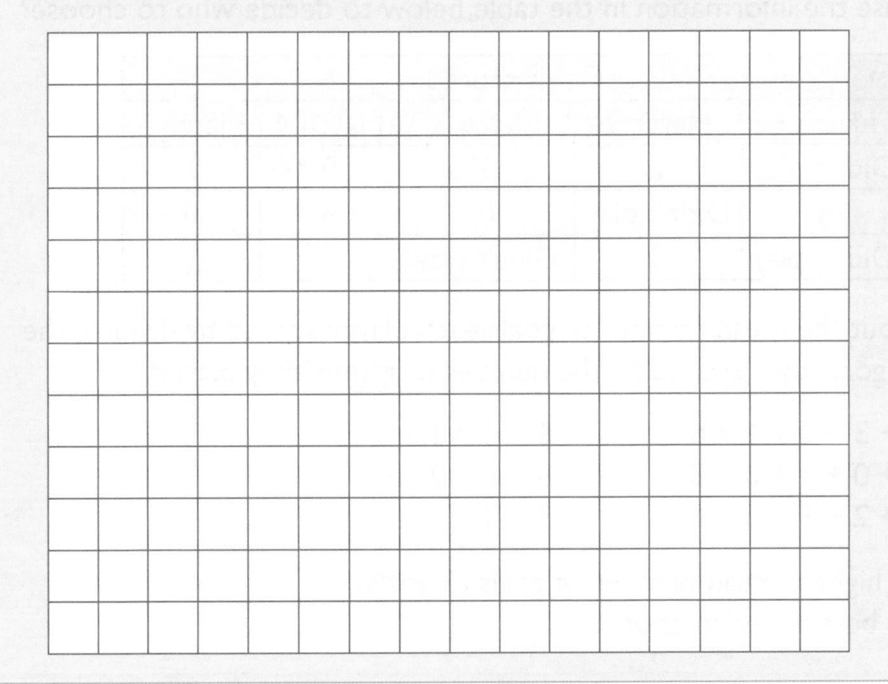

Year	Average Rent (£)
1990	250
1995	275
2000	300
2005	350
2010	475

Mean and Range

The Mean is a Type of Average

1) An average is a number that summarises a lot of data.
2) For example, the average national salary is calculated from the salaries of everyone in the country.

To work out the mean:

1) Add up all the numbers.
2) Divide the total by how many numbers there are.

EXAMPLE 1:

The table shows the times taken by a group of women to run 100 m. What is the mean time taken?

Runner	Time (s)
Jo	12.2
Rachel	13.6
Rebecca	11.9
Samantha	12.9
Catherine	14.0

1) First, add up the numbers:
12.2 + 13.6 + 11.9 + 12.9 + 14.0 = 64.6
2) There are 5 numbers so divide the total by 5:
64.6 ÷ 5 = 12.92
3) The mean is **12.92 seconds**.

EXAMPLE 2:

Dave needs a new goal scorer for his football team. He decides to move either Richard, Jeremy or Paul up from the B team.

How can Dave use the information in the table below to decide who to choose?

	Goals scored				
	Match 1	Match 2	Match 3	Match 4	Match 5
Richard	Didn't play	2	3	0	1
Jeremy	4	Didn't play	0	4	0
Paul	Didn't play	0	Didn't play	2	1

Dave can work out the mean number of goals each player scored by dividing the total number of goals they scored by the number of games they played.

Richard: 2 + 3 + 0 + 1 = 6   6 ÷ 4 = 1.5
Jeremy: 4 + 0 + 4 + 0 = 8   8 ÷ 4 = 2
Paul: 0 + 2 + 1 = 3   3 ÷ 3 = 1

Jeremy has the highest mean number of goals (2 goals), so looks like the best player to choose.

The Range is the Gap Between Biggest and Smallest

The range is the difference between the biggest value and the smallest value.

To work out the range:

1) Write down all the numbers in order from the smallest to the biggest.

2) Subtract the smallest number from the biggest number.

EXAMPLE:

The people waiting in a queue at a music festival are all asked their age.
The ages are: 18, 34, 18, 22, 20, 21, 26 and 24.
Work out the age range of the people waiting.

1) First, write the numbers in order of size:
18, 18, 20, 21, 22, 24, 26, 34.

2) The biggest number is 34 and the smallest is 18.

Range = 34 – 18 = **16 years.**

Practice Questions

1) Work out the means and ranges of the following sets of data.

a) 8, 12, 6, 15, 13, 5, 7, 10.

....

b) 23.1, 22.6, 23.7, 24.3, 26.2, 24.1.

....

2) David has been recording the temperature in his greenhouse over the course of a week. Here are the temperatures he recorded at different times of the day: 27 °C, 16 °C, 23 °C, 19 °C, 38 °C, 11 °C and 20 °C.

a) What was the mean and range of the temperatures that David recorded?

....

....

b) David missed off 37 °C in his list. Would including this temperature change the range?

....

Probability

Probability is all About Likelihood and Chance

1) Likelihood is how likely an event is to happen.

2) There are some key words you need to know:

   - Certain — this is when something will definitely happen.
     For example, getting a number between 1 and 6 when you roll a dice.
   - Likely — this is when something isn't certain, but there's a high chance it will happen.
     For example, it's likely that it will rain during the summer in the UK.
   - Even chance — this is when something is as likely to happen as it is not to happen.
     For example, there's an even chance of getting heads when you toss a coin.
   - Unlikely — this is when something isn't impossible, but it probably won't happen.
     For example, it's unlikely you'll win the jackpot in the lottery.
   - Impossible — this is when there's no chance at all of something happening.
     For example, it's impossible to roll a 7 on a standard six-sided dice.

3) An event being impossible isn't the same as one that is very very unlikely. For example, it's very very unlikely that it won't rain in the UK in winter, but it's not impossible.

Practice Questions

1) Describe the following events as certain, likely, even chance, unlikely or impossible.

   a) Your mother is younger than you.

   ....

   b) New Year's Day will be on the 1st of January next year.

   ....

   c) The weather will be warm and sunny on Christmas Day in the UK.

   ....

2) Gavin has 4 pairs of socks — 3 red pairs and one black. This morning he picked one to wear without looking. What is the likelihood of Gavin picking a red pair?

   ....

Test Help

Get to Know Your Chosen Method of Testing

1) There are two main ways that you could sit your test — either on paper, or using onscreen testing. The onscreen format could be unfamiliar to you.
2) For onscreen testing, there will be special tools for different question types, such as drawing charts, graphs and diagrams. You'll also use an onscreen calculator.
3) Look on the website of the exam board you're using — there may be sample questions that allow you to practise using these tools.

Always Show Your Working

1) In the test it's really important that you show all of your working — there are lots of marks for the methods you use and the calculations that you do.
2) If you don't show how you worked your answer out, you may not get all of the marks — even if your final answer is right.
3) So, even if you type a calculation into your calculator, you must also write the calculation down for the examiner to see. This applies to onscreen calculations too.

You May Have to Use an Answer in Another Calculation

1) Sometimes you may need to use the answer to one question to work out the answer to another question.
2) If you get the answer to the first question wrong, you'll also get the answer to the second one wrong.
3) BUT if you use the right method, and you use the answer that you got for the first question in your calculation, then you can still get full marks for the second question.
4) So even if you're unsure about an answer, don't give up — make sure you keep going until the end of the question.

Always Check Your Answers

It's really important that you check your answers. Checking your answers helps you to spot mistakes that you've made, and in some questions there are marks for showing that you've checked your answer. There are lots of ways you can check answers. For example...

1) Reverse the calculation (see pages 2 and 3 for more on this).
2) Do the calculation again using a different method to see if you get the same answer.
3) Think about whether your answer is sensible. For example, if your answer says that someone's lunch costs hundreds of pounds, you've probably made a mistake somewhere.

Task 1 — A Bike Ride

1. Helen and Nigel drive to a bike hire shop in Helen's car and park in the car park. There are two signs in the car park.

<u>Bike Hire Prices (per day)</u>
Adult Bike: £15
Child Bike: £10
A deposit of £50 per group is needed. It will be refunded when the bikes are returned.

(a) Helen pays for the car park, 2 adult bikes and the deposit. How much will this cost?

....

....

....

(2 marks)

(b) Helen and Nigel split the cost of the parking and the bikes equally between them. How much does Nigel owe Helen?

....

....

....

(2 marks)

(c) Show how you can check your answer to (b).

....

....

(1 mark)

2. Helen and Nigel use the map below while they are out on their bike ride.

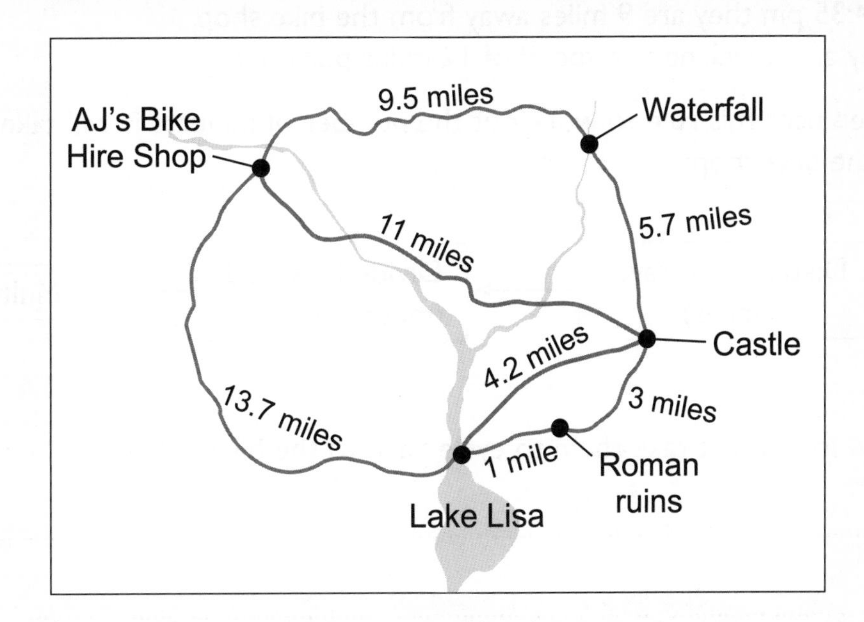

(a) Helen and Nigel cycle from AJ's Bike Hire Shop to the waterfall and then to the castle. How many miles is this?

....

....

....

(2 marks)

(b) Helen and Nigel are halfway between the Roman ruins and Lake Lisa when it starts to rain. What is the shortest route back to the bike shop from where they are now? You must show your working.

....

....

....

(2 marks)

3. Helen and Nigel need to return the bikes to the bike shop by 5:30 pm.

At 4:35 pm they are 9 miles away from the bike shop.
They are travelling at a speed of 12 miles per hour.

Helen uses this rule to work out the number of minutes it will take them to get back to the bike shop:

Distance to travel (miles) → Divide by speed (miles per hour) → Multiply by 60

(a) How long will it take them to cycle back to the bike shop?

....

....

....

(2 marks)

(b) Will Helen and Nigel get back to the bike shop in time? Give a reason for your answer.

....

....

....

(2 marks)

Task 2 — Shopping

4. Emma has gone shopping. She buys some food from the grocers. She checks her receipt because she thinks she has been overcharged.

```
chicken breasts
  2 @ £3.80 = £7.60
sandwich £2.20
crisps 45p
juice 60p
```

(a) The sandwich, crisps and juice Emma bought were on a £3 meal deal. How much has she been overcharged for these items?

....

....

(2 marks)

(b) The chicken breasts were on a 'buy one get one half price' offer. How much should Emma have been charged for the two chicken breasts?

....

....

....

(2 marks)

(c) How much should Emma get as a refund from the cashier?

....

....

....

(2 marks)

5. The owner of the shop gives Emma some vouchers to apologise for her being overcharged. The items Emma buys on her next shop are shown below:

bread £1.20
milk 80p
pasta £1.30
tea bags £2.40
cereal £2.20

How much will Emma save in total if she uses all the vouchers?

(5 marks)

Task 3 — Gardening

6. Aftab wants to make a patio in his garden. He is going to use concrete slabs.

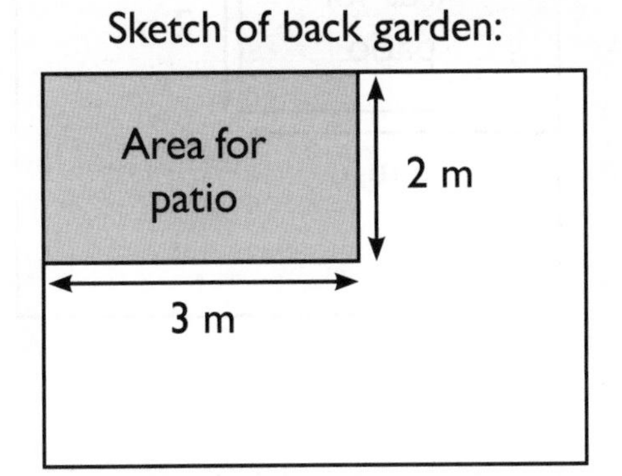

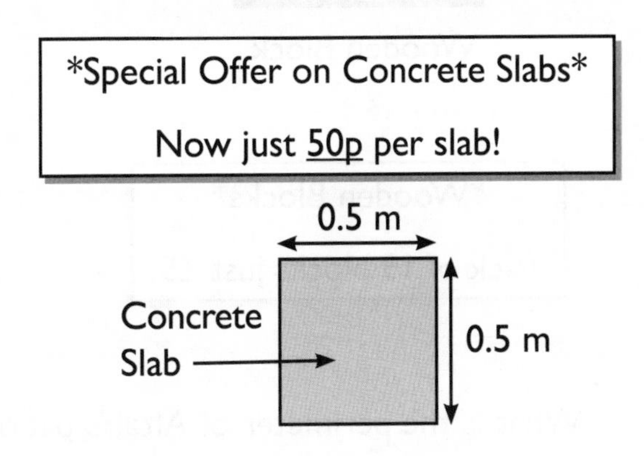

(a) What is the area of Aftab's patio?

....

....

(1 mark)

(b) How much will it cost Aftab to buy enough slabs to make the patio?

....

....

....

....

....

(3 marks)

7. Aftab decides to add a border made from wooden blocks all the way around his patio.

50 cm

Wooden Block

Wooden Blocks

Pack of 15 blocks just £5

Sketch of back garden:

Area for patio

2 m

3 m

(a) What is the perimeter of Aftab's patio?

....

....

(1 mark)

(b) How many wooden blocks will he need to buy to make the border?

....

....

....

(2 marks)

(c) How much will it cost to buy enough packs of wooden blocks?

....

....

....

(2 marks)

8. Helena has bought some liquid lawn feed to put on her lawn.
She reads the instructions to find out how much to use.

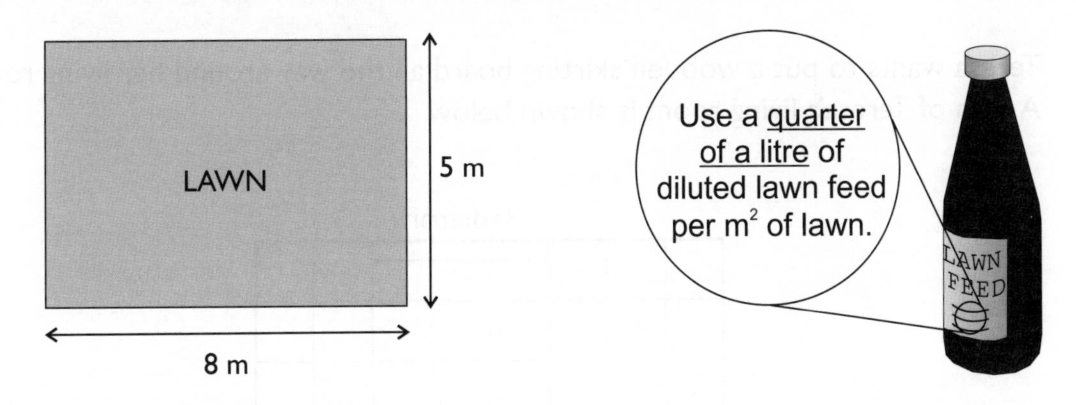

How much diluted lawn feed does Helena need to treat the lawn?

...

...

...

...

(3 marks)

9. Helena reads the instructions to find out how to dilute the lawn feed.

Helena measures out 7.5 litres of water in her watering can.
How much lawn feed does she need to add to this?

...

...

...

...

(3 marks)

Task 4 — Home Improvements

10. Teresa wants to put a wooden skirting board all the way around her living room walls. A plan of Teresa's living room is shown below.

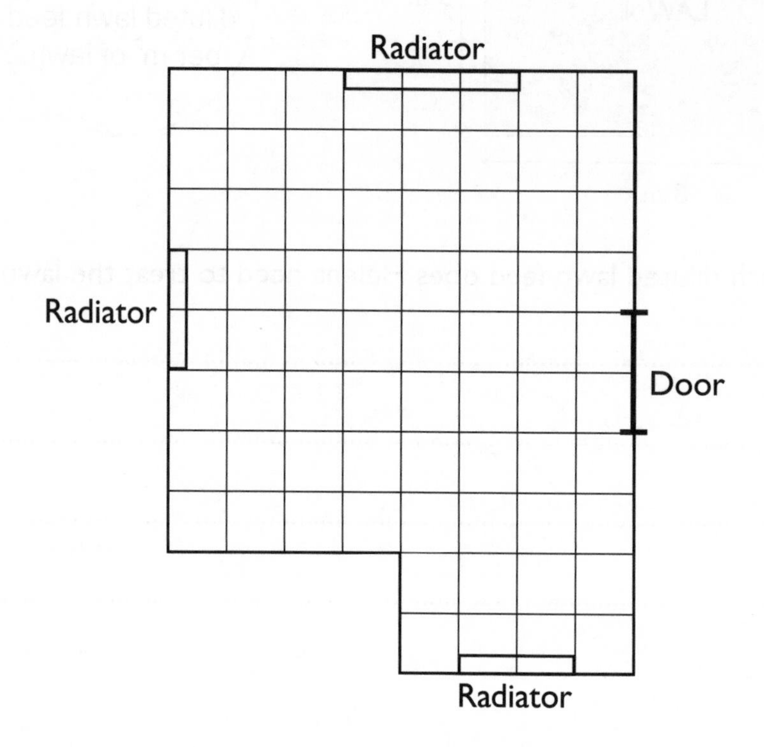

The length (and width) of each square on the plan is equal to 0.5 m.

(a) What is the total perimeter of the room?

....

....

(2 marks)

(b) Teresa needs the skirting board to go all the way around the walls, but not across the door. What length of board should she buy?

....

....

(2 marks)

11. Teresa has bought a sofa and a rug for her living room.
She wants to know where to put them.

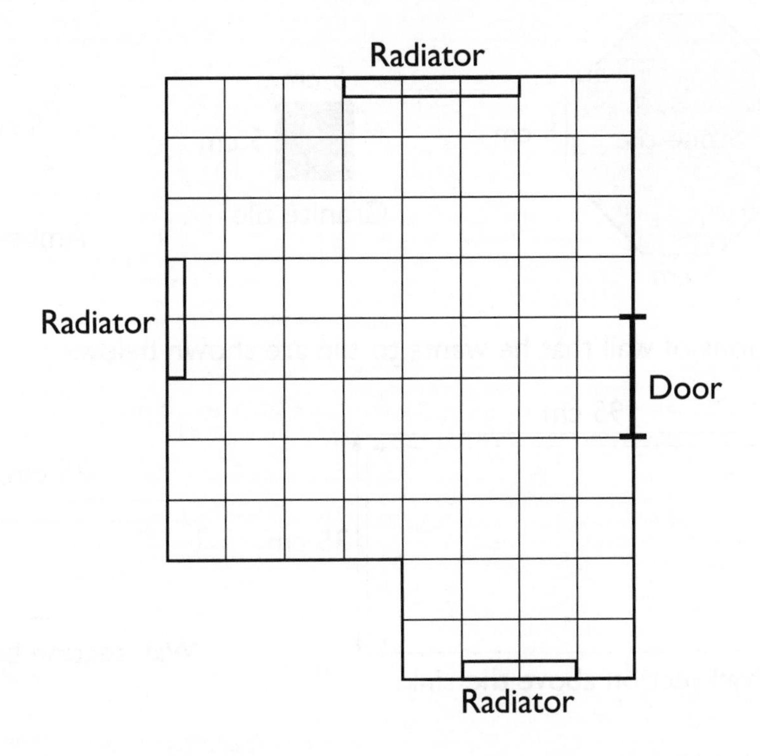

The length (and width) of each square on the plan is equal to 0.5 m.

- The sofa is 2 m long and 1 m wide.
- The sofa must not be against a radiator. It must be at least 1 m from the door.
- The rug is 1 m long and 0.5 m wide.
- The rug must be 0.5 m from the sofa.

On the plan above, draw where the sofa and the rug could go.

(4 marks)

12. Vin has three types of tiles (shown below). He wants to use them to tile two sections of wall in his bathroom. One section of wall is above the sink, the other is by the bath.

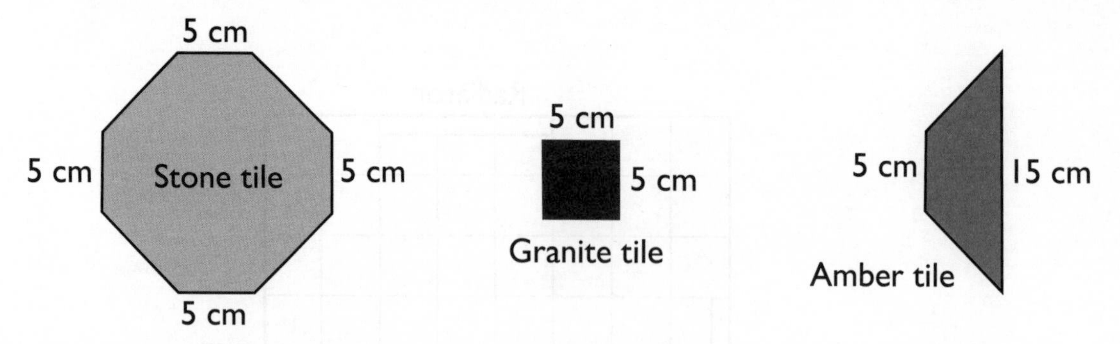

The sections of wall that he wants to tile are shown below.

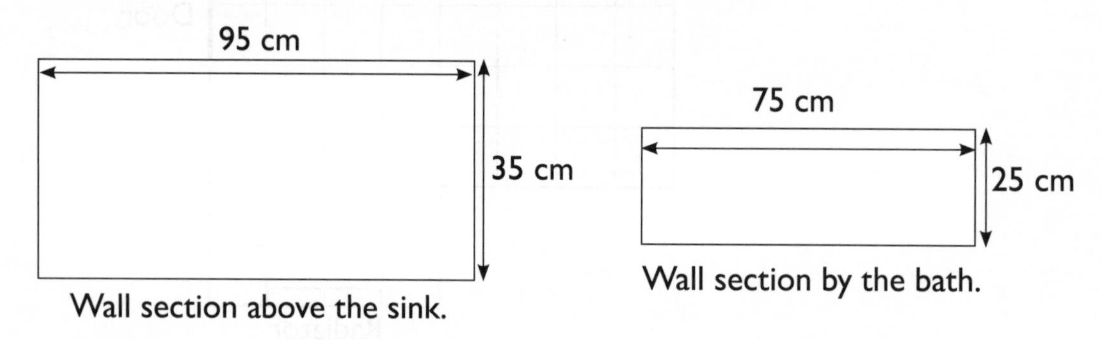

(a) Vin will use stone tiles and granite tiles to create a pattern on the wall above the sink. He doesn't want to leave any gaps between these tiles. He will only use amber tiles to fill in gaps at the edge of the wall.

Complete the diagram on the grid below to show how Vin could use stone, granite and amber tiles to completely cover the section of wall above the sink.

Scale: Length of 1 square on the grid = 5 cm on the wall.

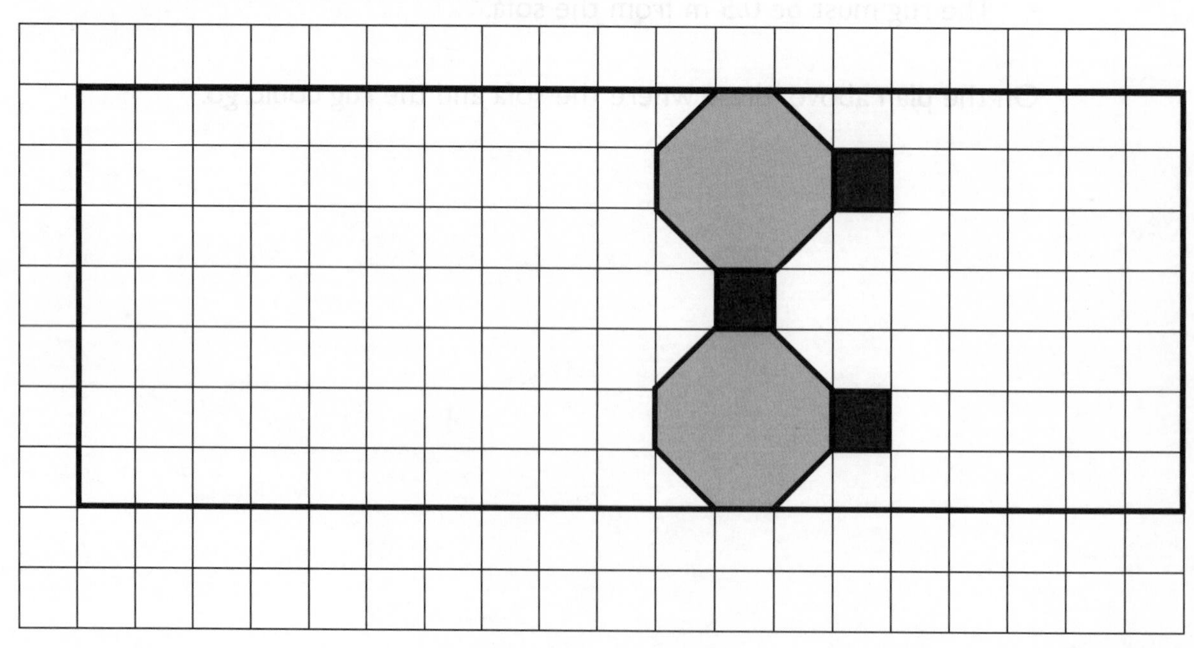

(2 marks)

(b) After tiling the section of wall above the sink, Vin will have 78 granite tiles left. He decides to use them to tile the area by the bath. Does he have enough tiles to cover this area?

....

....

....

....

....

(3 marks)

(c) Vin needs to buy some adhesive to fix the tiles to the wall. He needs to buy enough adhesive to cover both of the wall sections that he is tiling. He wants to spend as little money as possible. Use the table below to work out which tub(s) Vin should buy.

Tub size	Price	Area covered (m^2)
1 litre	£2.99	0.4
750 ml	£1.99	0.3
500 ml	£1.50	0.2

....

....

....

....

....

....

....

(5 marks)

Task 5 — A Trip to the Zoo

13. Mike is going to visit the zoo in Dayton with his friends Sophie, Michelle and Jacob.

They are all driving to the zoo from where they live. Mike lives in Millom, Sophie lives in Grizebeck, Michelle lives in Greenodd and Jacob lives in Barrow.

Millom				
11.6	Grizebeck			
17.8	6.3	Greenodd		
24.4	13	14.3	Barrow	
18.4	7.1	9.4	5.1	Dayton (the zoo)

Distances are in miles.

(a) Who has to travel the furthest to get to the zoo?

(1 mark)

(b) Is the zoo closer to where Sophie lives or where Michelle lives?

(1 mark)

(c) Mike's car costs him £0.60 per mile to run.

How much will the journey from his house to the zoo cost him?

(2 marks)

(d) Michelle's car has broken down, so Mike offers to give her a lift. This means Mike will have to travel from Millom to Greenodd and then to the zoo.

How much more will Mike's journey cost when he gives Michelle a lift?

(3 marks)

14. Mike and his friends are working out the cost of tickets for the zoo.

Admission Prices

Adult: £13.00
Child: £8.00
Seniors and Students: £9.50

(a) Jacob and Michelle are students. Mike and Sophie are not students and have to pay the normal adult price. How much will it cost for all four of them to enter the zoo?

....

....

(1 mark)

(b) Mike remembers that he has a voucher for 10% off when you buy four adult tickets.

Is buying four adult tickets using the voucher cheaper than buying two adult and two student tickets at normal price? You must show how you got your answer.

....

....

....

(3 marks)

(c) For £35 you can buy an annual pass for the zoo. With a pass you can come back as many times as you like for a year without having to pay each time.

Mike would normally pay one full adult price each visit. If Mike visits three times this year will it be worth it for him to buy an annual pass? Explain your answer.

....

....

....

(2 marks)

15. At the zoo there is a list of the feeding times for different animals.

Daily Feeding Times

Giraffes — 12.00
Lions — 10.30 and 13.30
Penguins — 11.30
Vultures — 10.30 and 14.00
Monkeys — 15.00
Lemurs — 13.30

Each feeding session lasts for about 20 minutes.

(a) What order should the group visit the animals in so that they see them all being fed? Draw up a time plan to show the order of visits.

(2 marks)

(b) The group would like to have a one hour lunch break. When can they fit this in without missing any feeding times? Explain your answer.

....

....

(2 marks)

(c) At 12.00 they find out the lemur feeding has been moved to 14.00. Can the group still see all the animal feeding sessions? Explain your answer.

....

....

(1 mark)

16. At lunchtime the group go to a restaurant at the zoo.
They notice there is a meal deal available.

Sandwiches £2.40
Salads £2.70
Drinks £1.30
Crisps £0.50
Cakes £0.55

Special offer: Adult meal deal for £3.50

Meal deal includes sandwich, drink and packet of crisps.

(a) Mike, Michelle and Jacob decide to buy a meal deal. How much will they each save over buying all the items included in the meal deal at their normal price?

....

....

(2 marks)

(b) Sophie would prefer a cake instead of crisps, so decides to buy a sandwich, drink and cake separately. How much more will her lunch cost than the meal deal?

....

....

(2 marks)

(c) Jacob thinks that Sophie can buy a meal deal plus a cake and it will be cheaper than buying just a sandwich, drink and cake separately. Is he right? If so, how much cheaper will it be?

....

....

....

(3 marks)

Task 6 — Summer Holiday

17.(a) Mr and Mrs Landy and their two children are planning to go on holiday next year. They want to go for one week in August.

The table below shows combined prices for flights and hotel.

Number of Nights:	7		14	
Board:	Half	Full	Half	Full
01 Jan - 28 Feb	140	190	200	290
01 Mar - 30 Apr	180	240	250	330
01 May - 30 Jun	270	300	360	450
01 Jul - 31 Aug	330	400	520	630
01 Sep - 31 Oct	230	270	310	420
01 Nov - 31 Dec	270	310	380	470

Prices shown are in **£s per adult**. Child prices are **half** the adult price.

(i) How much will it cost for the family to go half-board?

....

....

....

(2 marks)

(ii) The Landys have a budget of £1100.
If they go half-board how much of their budget will the Landys have left?

....

....

(1 mark)

The family are offered a special discount. If they book to go in June they can get two weeks full-board for the same price as two weeks half-board.

(iii) Can they afford the discounted June holiday with their £1100 budget?

....

....

....

(2 marks)

(b) The Landy family are getting the train from Chorley to Manchester Airport for their flight.

It takes 15 minutes to walk to Chorley station from the Landy's house, and they will need 5 minutes at the station to collect their train tickets.

Trains to Manchester Airport

Lancaster	0747	0926	—	1126
Preston	0807	0945	1004	1145
Chorley	0822	0956	1022	1156
Bolton	0834	1008	1034	1208
Manchester Oxford Road	0852	1023	1052	1223
Manchester Piccadilly	0856	1027	1056	1227
Manchester Airport	0919	1047	1117	1247

Flight Details
Flight No.: BNT3891

Departure Time: 1400

Please arrive no later than 2 hours before departure

(i) What is the latest train from Chorley that the Landys can catch to arrive in time for their flight? Explain how you got your answer.

....

....

(2 marks)

(ii) What is the latest time that the Landys can leave their house to arrive in time for their train? Explain how you got your answer.

....

....

(2 marks)

(iii) The day before they are due to catch their flight the Landys get an email saying that their flight has been rescheduled. It will now depart at 14:30 instead of 14:00.

If the Landys catch the 11:56 from Chorley, would they arrive at the airport early enough to catch their flight? Explain your answer.

....

....

....

(3 marks)

(c) The Landy family has arrived at the airport.
The airline they are flying with has a baggage limit of 15 kg per person.

Weight in stones	Weight in kilograms
1.7	10.8
1.8	11.4
1.9	12.1
2.0	12.7
2.1	13.3
2.2	14.0
2.3	14.6
2.4	15.2
2.5	15.9
2.6	16.5
2.7	17.1
2.8	17.8

Baggage weight limit

15 kg per person

At home the Landys weighed their bags on scales that showed the weight in stones.

Mr Landy's bag weighed 2.8 stones.
Mrs Landy's bag weighed 2.3 stones.
Their daughter's bag weighed 1.7 stones.
Their son's bag weighed 2.0 stones.

The table above can be used to convert between stones and kilograms.

(i) By how many kilograms is Mr Landy's bag over the 15 kg limit?

....

....

(2 marks)

(ii) Mr Landy thinks he can transfer the excess weight to another family member's bag.

Whose bag could he add the excess weight to without taking the bag over the weight limit? Show how you get your answer.

....

....

....

....

....

(3 marks)

(d) The Landys have arrived at their hotel. They are planning a boat trip for their first day. They have found a few adverts for boat trips in the hotel lobby. The prices for the boat trips are shown in euros (€).

Advert 1

Boat Trips

Every day from the harbour

Adults — €15 each
Children — €7.50 each

Advert 2

1 Hour Boat Tours
Leaving daily at 10am
from the harbour

Adult: €13
Child: €7
Family Pass €35
(2 adults and
up to 3 children)

Advert 3

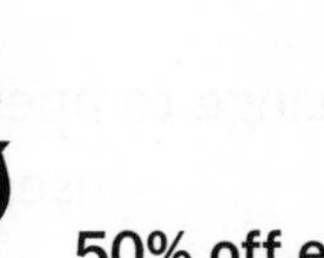

90 Minute Cruise
Adult: €16
Child: €10
Senior: €8
50% off each child's ticket when bought with an adult ticket

(i) How much does the family pass in Advert 2 save over buying two adult and two child tickets at their normal prices?

....

....

(2 marks)

(ii) How much does the special offer in Advert 3 save over buying two adult and two child tickets at their normal prices?

....

....

(2 marks)

(iii) Which boat tour would work out the cheapest for the family to go on?

....

....

....

....

....

(4 marks)

Task 7 — In the Kitchen

18.(a) Katie has been given an old recipe for a cake she wants to bake.

Her electronic scales and measuring jug only show millilitres and grams, so she needs to convert the amounts shown on the old recipe.

<u>Cake recipe</u>

1 lb butter

1 ½ lb sugar

2 lb flour

6 eggs

½ pint of milk

Set oven to 330 °F...

<u>Conversion instructions</u>

1 lb is about 450 g

1 pint is about 570 ml

To change temperatures from °F to °C, use this rule:

"Take away 32, then multiply by 5. Divide your answer by 9."

(i) Work out the amounts of butter, sugar and flour Katie needs in grams.

....

....

....

(3 marks)

(ii) How much milk does she need in ml?

....

....

(1 mark)

(iii) Katie's oven is marked in °C. What temperature should she set it to?

....

....

....

(3 marks)

(b) Katie wants to make some fairy cakes for her friends.
She wants to make 12 cakes from the recipe shown below.

<u>Fairy Cakes</u>
(Makes 36)

375 g butter

750 g sugar

600 g flour

6 eggs

360 ml milk

(i) Work out the amounts of all the ingredients Katie needs to make 12 cakes.

....

....

....

....

....

....

(3 marks)

(ii) If Katie used this recipe to make 48 cakes, how much sugar and flour would she need?

....

....

....

....

(2 marks)

(c) Harinder is making a small fruit cake. She needs 6½ oz flour and 4½ oz brown sugar. She weighs the flour out in a bowl and starts adding sugar to the same bowl.

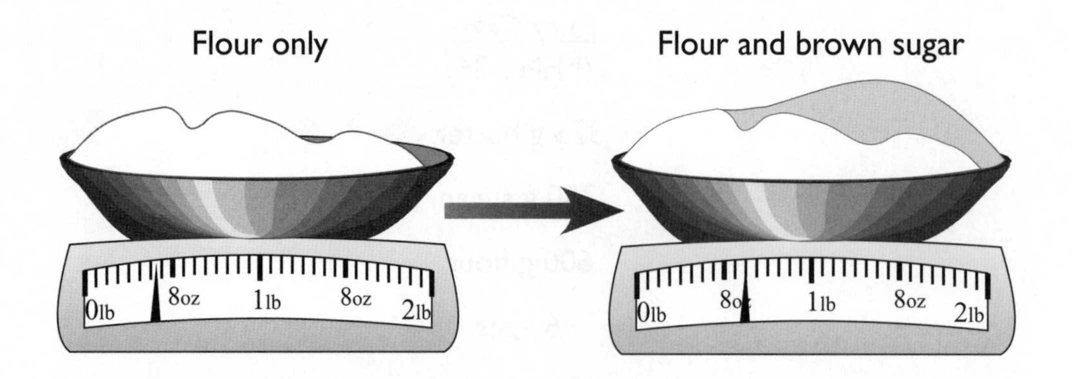

(i) Has Harinder added the correct amount of brown sugar? If not, how much more or less does she need?

....

....

....

(1 mark)

(ii) Harinder also needs to add 4 oz of butter. What should the scale read with the right amounts of flour, sugar and butter in the bowl together?

....

....

(1 mark)

(iii) Harinder finds another fruit cake recipe. In the recipe the ratio of butter to flour is 1 : 4. If she uses 3½ oz of butter, how much flour will she use?

....

....

(2 marks)

(d) Maria is measuring out ingredients to make carrot and lentil soup using weighing scales and a jug. These are shown in diagrams A and B below. The soup recipe is also shown below.

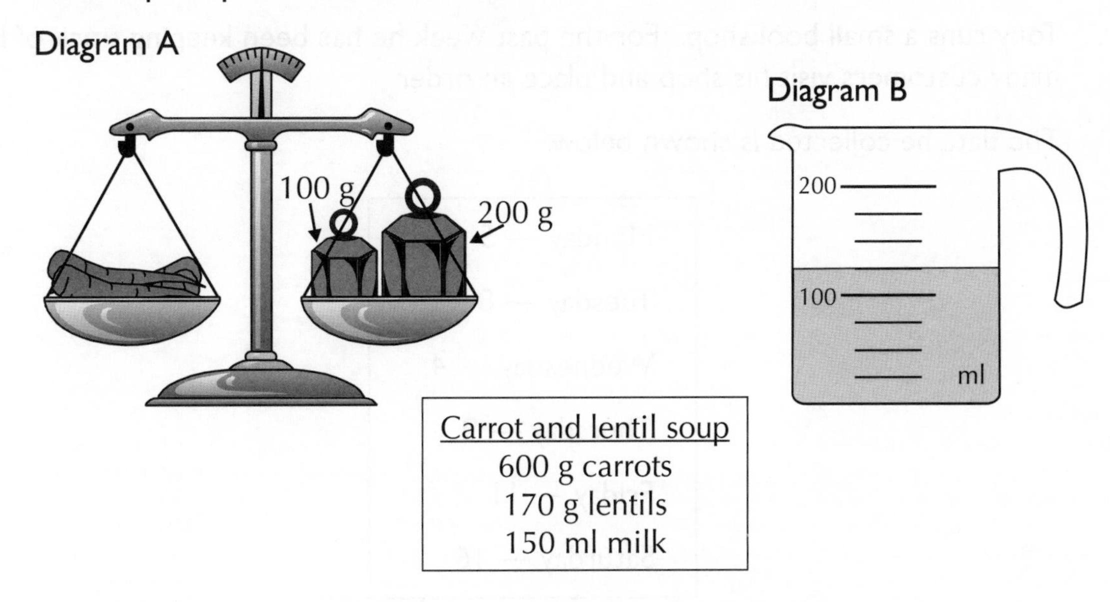

Maria's scales come with one each of the following weights:

200 g, 100 g, 50 g, 20 g, 5 g and **1 g.**

(i) What combination of weights should Maria use on her scales to weigh out the lentils?

....

....

(1 mark)

(ii) Maria has put four carrots on her scales, as shown in diagram A above. Roughly how many more carrots will she need to make the soup?

....

....

....

(3 marks)

(iii) Maria has poured some of the milk for the soup into a measuring jug, as shown in diagram B above. How much more milk does she need to add to the jug?

....

....

(2 marks)

Task 8 — Tony's Bookshop

19.(a) Tony runs a small bookshop. For the past week he has been keeping track of how many customers visit his shop and place an order.

The data he collected is shown below.

Monday — 5

Tuesday — 8

Wednesday — 4

Thursday — 9

Friday — 11

Saturday — 16

Draw a tally chart to show the data Tony collected.

(3 marks)

(b) Tony has been collecting information about how well children's books are selling. He has created a table showing how many children's books he has sold each week for the past 6 weeks.

Week number	Books sold
1	10
2	11
3	8
4	13
5	16
6	20

(i) Work out the mean number of children's books sold in a week.

....

....

(2 marks)

(ii) Work out the range for the number of children's books sold.

....

....

(2 marks)

(iii) Draw a graph or chart to show the information in the table.

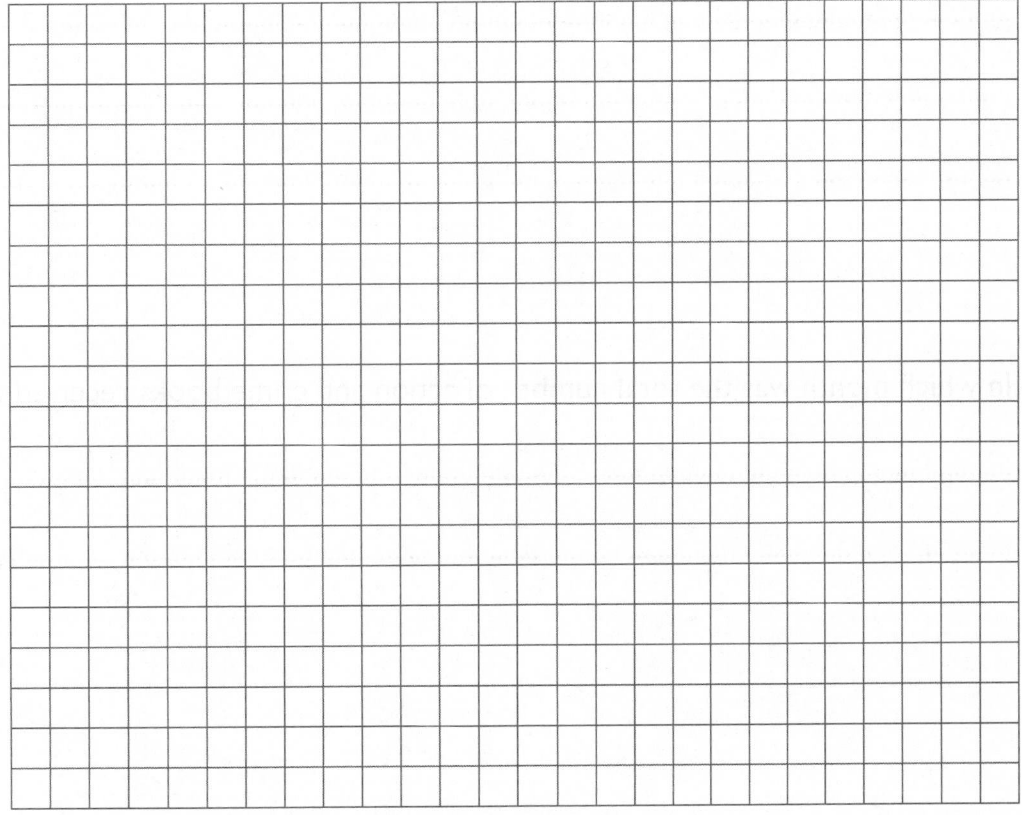

(4 marks)

(c) Tony receives a delivery of newly released fiction books once a month. For June and July he has been keeping track of the categories that they fall into.

June

Category	Number of books
Action	7
Crime	8
Romance	3
Historical	7
Sci-Fi	4
Horror	4

July

Category	Number of books
Action	5
Crime	9
Romance	7
Historical	2
Sci-Fi	5
Horror	2

(i) In which month did Tony receive the most new books?

....

....

....

(2 marks)

(ii) From which category did Tony receive the most new books in June and July in total?

....

....

....

....

(2 marks)

(iii) In which month was the total number of action and crime books received the highest?

....

....

(2 marks)

Answers — Practice Questions

Section One — Number

Page 2
Q1 £9
Q2 £4
Q3 11
Q4 £154
To check it: 154 + 346 = 500

Page 4
Q1 45
Q2 8
Q3 363 ÷ 11 = 33 or
363 ÷ 33 = 11
Q4 a) 12
b) 54
Q5 Yes (it will have 924 rulers).
Q6 a) 400
b) 1200

Page 5
Q1 8
Q2 23
Q3 3

Page 7
Q1 0.4
Q2 2
Q3 3
Q4 72 hours

Page 9
Q1 -8
Q2 At night
Q3 23 °C

Page 10
Q1 -1 °C
Q2 -5 °C

Page 11
Q1 $\frac{3}{5}$

Q2 $2\frac{1}{2}$

Page 12
Q1 6
Q2 20
Q3 a) 10.5 hours (or 10½ hours)
b) 14 hours

Page 13
Q1 34p
Q2 £380
Q3 £90

Page 15
Q1 0.03 kg, 0.6 kg, 1.1 kg
Q2 1.85 cm
Q3 No. He has only spent £19.56.

Page 17
Q1 3.6
Q2 1.02
Q3 15.9 km
Q4 £181.75
Q5 3.6 litres
Q6 £41.68
In questions like these last few, you often need to include units in your answer. For example, £ or km. If you're not told what units to use, stick with the ones in the question.

Page 19
Q1 $\frac{34}{100}$ (or $\frac{17}{50}$)
Q2 4.8
Q3 11.25 (or 11¼)
Q4 a) 16
b) 13
Q5 £3.50

Page 20
Q1 a) £640
b) £2560
Q2 2.75 m
Q3 £5450

Page 22
Q1 0.5
Q2 $\frac{1}{4}$
Q3 1.25
Q4 a) 60%
b) 0.6
Q5 a) 40%
b) Neither / They both have the same percentage of people responding (40%).

Page 23
Q1 0.25
Q2 Darren. He pays £360.
Leanne pays £400.

Page 25
Q1 a) 1:4
b) 100 ml
Q2 18
Q3 The first person gets £2000.
The second person gets £1000.

Page 27
Q1 48 kg
Q2 750 ml
Q3 a) 4
b) 360 g
Q4 £1500

Page 30
Q1 £56.25
Q2 a) 120 minutes (or 2 hours)
b) 90 minutes (or 1.5 hours)
Q3 £174
Q4 a) 32.5 m^2 (or 32½ m^2)
b) Yes. 32 m^2 is less than 32.5 m^2.

Section Two — Measure

Page 33
Q1 100
Q2 1.3 kg
Q3 50 ml
Q4 5 km

Page 34
Q1 12.5 cm
Q2 2.55 kg
Q3 £11.20
Q4 360 Calories

Page 36
Q1 1066 g
Q2 105 kg
Q3 9
Q4 Yes, David and his equipment weigh a total of 56.4 kg.
Q5 6
Q6 200
Q7 2

Page 38
Q1 a) 300 ml
b) 8 ml
c) 150 ml
Q2 5
Q3 Yes, the measurements add up to 135 ml which is less than 150 ml.

Page 39
Q1 a) 18 cm
b) 26 cm

Page 41
Q1 a) 24 m
b) 12 cm
Q2 a) 24 cm (the unknown side is 4 cm long).
b) 28 cm (the unknown sides are 4 cm and 3 cm long).
c) 24 cm (the unknown sides are 4 cm and 2 cm long).
Q3 13.25 m

Page 43
Q1 a) 6 cm^2
b) 42 cm^2
c) 30 cm^2
d) 32 cm^2

Page 45
Q1 £693
Q2 96 tiles

Page 47
Q1 a) 12 cm^3
b) 10 cm^3
Q2 a) 24 cm^3
b) 50 cm^3
Q3 0.16 m^3

Page 49
Q1 a) 127p
b) £2.19
Q2 The 6-pack is the best value for money. (£0.20 per pack. The price per packet of the 10-pack is £0.23.)
Q3 The 50 g tin is the best value. (It costs £1.26 per gram. The 30 g tin costs £1.30 per gram.)

Page 51
Q1 120 minutes
Q2 30 years
Q3 4 minutes
Q4 a) 9:00 am
b) 4:45 pm
Q5 a) 17:15
b) 07:05
Q6 No (10:47 pm is 22:47).

Page 53
Q1 2 hours 15 mins (or 135 mins)
Q2 2 hours 52 minutes (or 172 mins)
Q3 7:50 pm (or 19:50)
Q4 1 hour 55 minutes (or 115 mins)
Q5 20:15 (or 8:15 pm)
Q6 Yes (if it takes a quarter of an hour to find a parking space she should get to the theatre at 19:15).

Page 56
Q1 a) 1900
b) 1845
Q2 14:00 on Tuesday, or 13:00 on Thursday, or 11:00 on Friday.
Q3 E.g.

	Day 1
10:00	K Craig
11:25	S Williams
12:05	H Gregson
12:30	Lunch
13:30	M Hamill

	Day 2
10:00	M Falkner
12:10	J Towle
12:55	Lunch
13:55	M Tyler

Other answers and layouts are possible.

Section Three — Shape and Space

Page 58
Q1 90° = A
180° = C
270° = B
Q2 a) 80°
b) 140°
c) 170°
Q3 180°

Page 60
Q1 a)

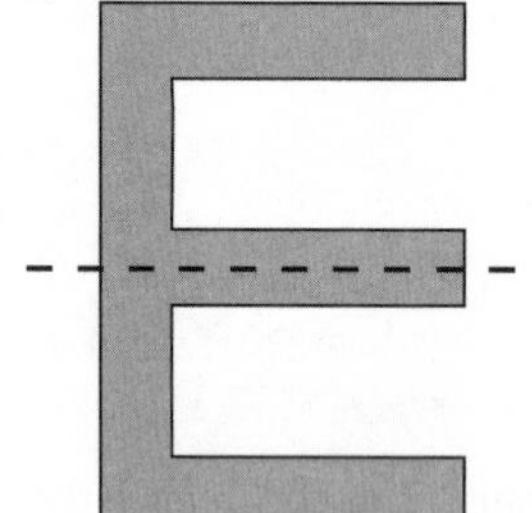

b)

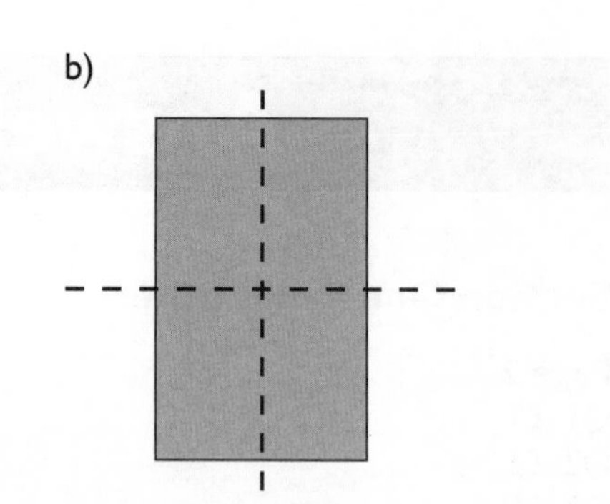

c)

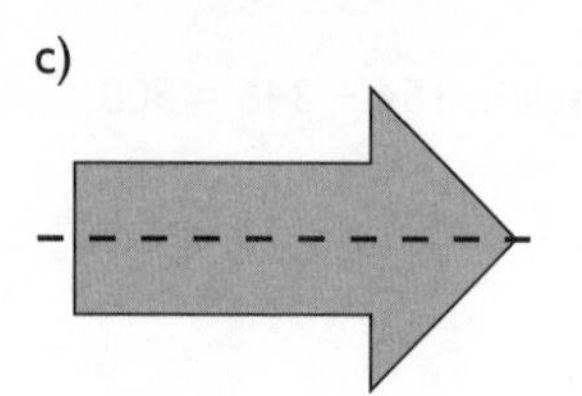

Q2 a)
b)

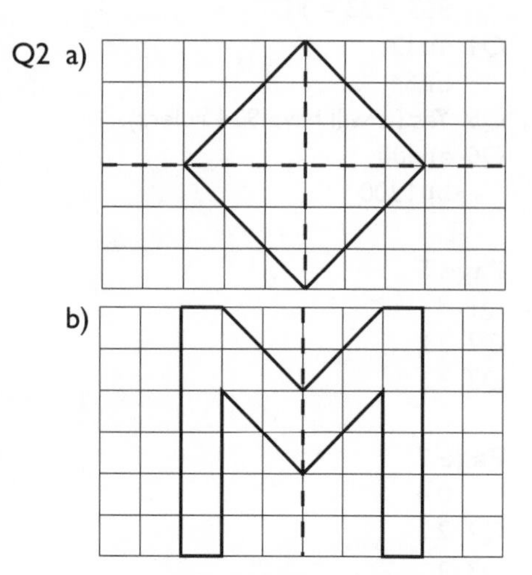

Q3 a)
b)

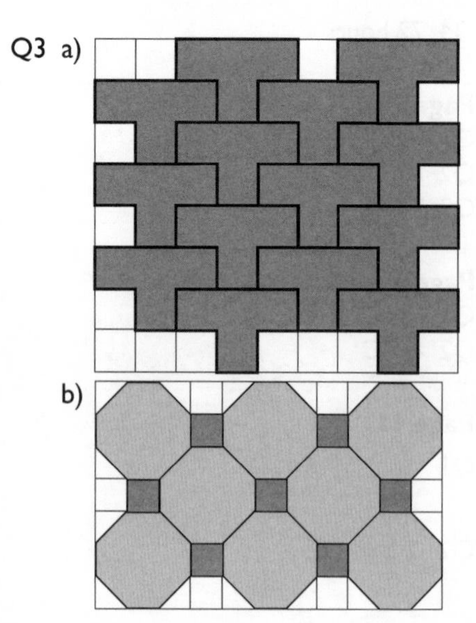

Page 63

Q1 a) Yes

b)

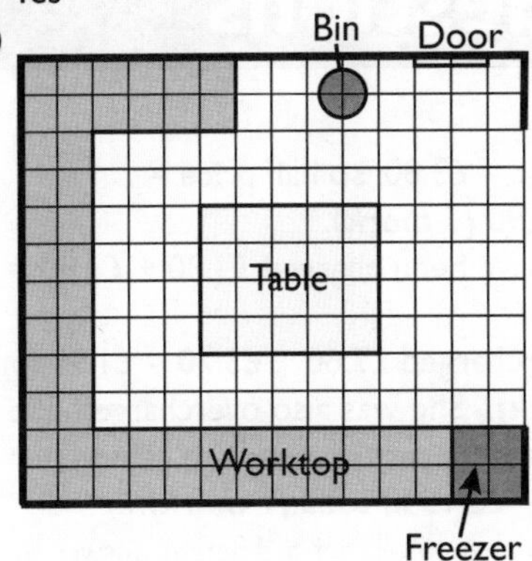

Q2 a) Climbing frame (see right) can be placed anywhere in the highlighted area.

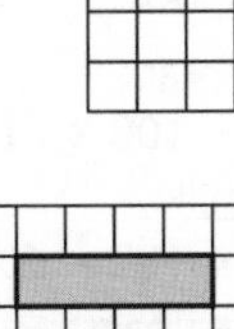

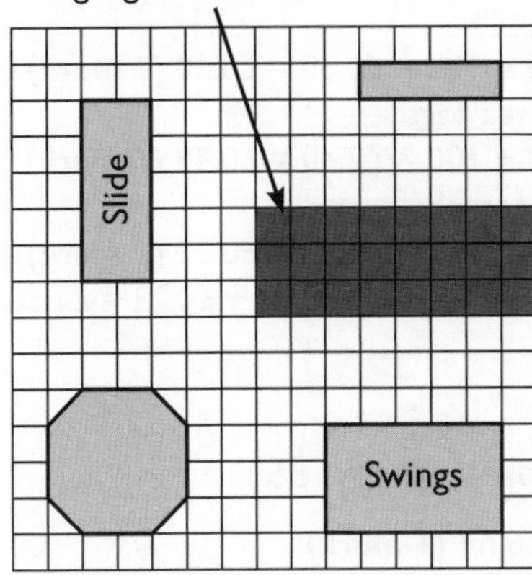

b) No. The climbing frame would be less than 3 m from the swings and the see-saw.

Section Four — Handling Data

Pages 66-67

Q1 a) 1.4 litres
b) 14.1 seconds
c) £9867

Q2 a) 292 miles
b) 456 miles
c) 458 miles

Q3 a)

Type of tree	Tally	Frequency
Beech	III	3
Oak	II	2
Yew	III	3
Cedar	~~IIII~~	5
Pine	I	1
		Total 14

b) 5
c) 2

Q4 E.g.

Drink	Saturday	Sunday
Tea	75	60
Coffee	60	41
Fizzy	86	59
	Total 221	Total 160

You might have drawn a table that looks different to this one. As long as it shows how many of each type of drink were sold on each day, and the total number of drinks sold on each day, then it's correct.

Page 68

Q1 a) 3
b) Yellow
c) 15

Page 70

Q1 a) 1400 litres
b) 900 litres
c) 500 litres

Q2 a) 45 mm
b) 20 mm
c) 4 weeks

Q3 a) $8
b) £7.50
c) £8

Page 72

Q1 a) Yoga
b) 25%
c) Under 25 years old

Q2 a) i) 100 pizzas
ii) 80 pizzas
iii) 40 pizzas
b) 220 pizzas

Page 75

Q1 a) 250

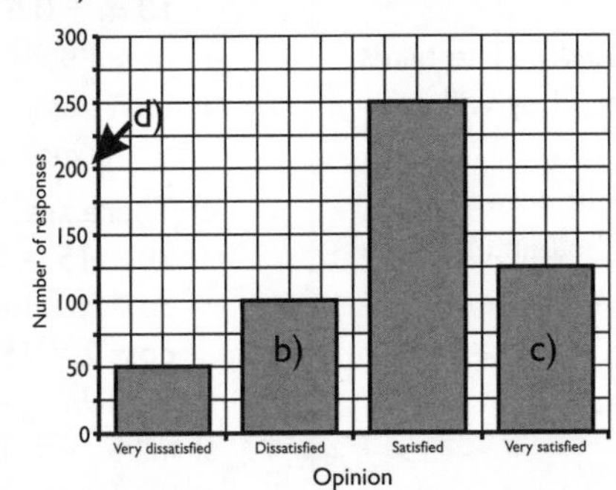

Q2 For example:

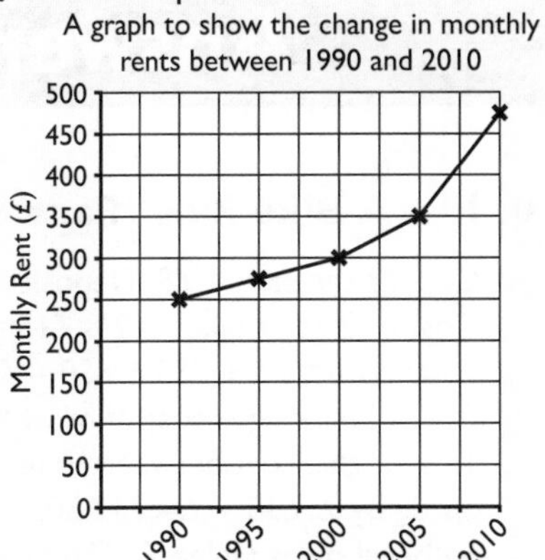

Your graph may not look exactly like the one above. For example, you might have used a different scale for your axes.

Page 77

Q1 a) Mean = 9.5, Range = 10
b) Mean = 24, Range = 3.6

Q2 a) Mean = 22 °C
Range = 27 °C
b) No (it's between 11 and 38°C)

Page 78

Q1 a) Impossible
b) Certain
c) Unlikely

Q2 Likely (OR ¾ or 75%).

Answers — Test-Style Questions

Task 1 — A Bike Ride (Page 80)

1 a) Cost of car park = £5 Deposit = £50
Cost of 2 adult bikes = 2 × £15 = £30 ***(1 mark)***
Total cost = £5 + £30 + £50 = £85 ***(1 mark)***

b) Cost of parking + cost of bikes = £5 + £30 = £35.
To split this cost between Helen and Nigel you need to divide by 2: £35 ÷ 2 = £17.50.
So Nigel owes Helen £17.50 ***(1 mark for attempting to divide total cost of bikes and parking by 2, 1 mark for £17.50).***

c) E.g. Helen and Nigel pay the same amount for the bikes and parking, and then Helen pays an additional £50 for the deposit. The amount that they both pay for the bikes, plus £50, should add up to the total amount (£85).
£17.50 + £17.50 + £50 = £85
(1 mark for any correct check).

There are a few different ways you can check your answer to b). If your working is correct you'll get the mark.

2 a) Distance to the waterfall = 9.5 miles.
Distance from the waterfall to the castle = 5.7 miles.
Total distance = 9.5 + 5.7 = 15.2 miles ***(1 mark for correct working, 1 mark for correct answer).***

b) The distance from the Roman ruins to Lake Lisa is 1 mile, so Helen and Nigel are half a mile from both.
Possible routes back are:
Lake Lisa, then straight back: 0.5 + 13.7 = 14.2 miles.
Roman ruins, the castle, then straight back:
0.5 + 3 + 11 = 14.5 miles.
Lake Lisa, then the castle, then straight back:
0.5 + 4.2 + 11 = 15.7 miles.
So the shortest route back is via Lake Lisa, and then straight back ***(1 mark for comparing at least two routes, 1 mark for correct answer).***

3 a) Distance to travel = 9 miles.
Speed = 12 mph
So it will take them 9 ÷ 12 × 60 = 45 minutes to get back to the shop ***(1 mark for correct working, 1 mark for correct answer).***

b) 45 mins after 4:35 pm is 5:20 pm ***(1 mark).*** The bikes need to be returned by 5:30 pm, so yes, they will get back in time ***(1 mark).***

You can get full marks for this part even if your answer to a) is wrong. As long as you've used the right method and used the answer that you got for a), then you'll get the marks.

Task 2 — Shopping (Page 83)

4 a) Change all the prices into the same units.
E.g. 45p = £0.45. 60p = £0.60.
Emma paid £2.20 + £0.45 + £0.60 = £3.25 ***(1 mark).***
She should only have paid £3 so she was overcharged by £0.25 (25p) ***(1 mark).***

b) 1 chicken breast = £3.80, so half price =
£3.80 ÷ 2 = £1.90 ***(1 mark).***
Emma should have been charged £3.80 + £1.90 = £5.70 ***(1 mark).***

c) Emma was overcharged £7.60 − £5.70 = £1.90 for the chicken ***(1 mark).*** She was also overcharged 25p for the meal deal, so the cashier needs to refund her £1.90 + £0.25 = £2.15 in total ***(1 mark).***

If you got b) wrong then you'll get a different answer to part c). But as long as you've used the right method for c), you should still get the marks.

5 <u>Amount saved on bread</u>:
50% of £1.20 = 50 ÷ 100 × £1.20 = £0.60 ***(1 mark).***
<u>Amount saved on milk</u>:
25% of £0.80 = 25 ÷ 100 × £0.80 = £0.20 ***(1 mark).***
<u>Amount saved on tea bags</u>:
20% of £2.40 = 20 ÷ 100 × £2.40 = £0.48 ***(1 mark).***
<u>Amount saved on cereal</u>:
10% of £2.20 = 10 ÷ 100 × £2.20 = £0.22 ***(1 mark).***
<u>Total saved</u> = £0.60 + £0.20 + £0.48 + £0.22 = £1.50 ***(1 mark).***

Task 3 — Gardening (Page 85)

6 a) Area is 3 × 2 = 6 m^2 ***(1 mark).***

b) E.g. area of patio = 6 m^2
Area of 1 slab = 0.5 m × 0.5 m = 0.25 m^2
Number of slabs need = 6 ÷ 0.25 = 24 slabs
(1 mark for any correct method, 1 mark for 24).

There's more than one way you can work out how many slabs are needed. As long as the method you've used is correct, you'll get the mark.

Each slab costs 50p, so the total cost of the slabs is 24 × £0.50 = £12 ***(1 mark).***

7 a) Perimeter is 3 + 2 + 3 + 2 = 10 m ***(1 mark).***

b) Each block is 50 cm long.
To convert cm to m, divide by 100:
50 ÷ 100 = 0.5 m ***(1 mark).***
10 m ÷ 0.5 m = 20 wooden blocks ***(1 mark).***

You'll get both marks for b) if you've used the right method and have used the answer that you got for a) — it doesn't matter if your answer to a) was wrong.

c) There are 15 blocks in each pack.
20 ÷ 15 = 1.3333
So Aftab needs to buy 2 packs ***(1 mark)***, which will cost 2 × £5 = £10 ***(1 mark).***

If you didn't get 20 as the answer to b) then your answer to c) will be different. You'll still get the marks, as long as your method is correct.

8 Area of the lawn: 5 × 8 = 40 m^2 ***(1 mark).***
A quarter of a litre = 0.25 litres, so Helena needs 0.25 litres of lawn feed to treat 1 m^2 of lawn ***(1 mark).***
So for 40 m^2 of lawn, she will need:
0.25 × 40 = 10 litres of lawn feed ***(1 mark).***

9 There are 7.5 L of water in the watering can.
To convert L to ml, multiply by 1000:
7.5 × 1000 = 7500 ml ***(1 mark)***.
This is 100 times as much water as 75 ml.
So Helena will need 100 times as much lawn feed:
5 × 100 = 500 ml ***(1 mark for any correct method, 1 mark for 500 ml)***.

Task 4 — Home Improvements (Page 88)

10 a) Each square = 0.5 m, so two squares = 1 m.
Perimeter is 4 m + 4 m + 2 m + 1 m + 2 m + 5 m = 18 m. ***(1 mark for correct working, 1 mark for correct answer)***

b) The distance around the walls is 18 m, the width of the door is 1 m, so she needs:
18 – 1 = 17 m. ***(1 mark for correct working, 1 mark for correct answer.)***

If you didn't get the right answer to a) then your answer for b) will be different. You can still get both marks, as long as you use the right method, and use the answer that you got for a).

11 E.g.

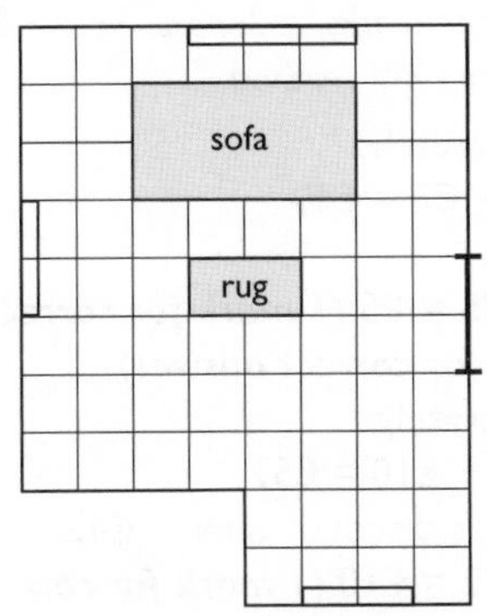

Other layouts are possible.
(1 mark for a correctly sized sofa, 1 mark for a correctly positioned sofa, 1 mark for a correctly sized rug, 1 mark for a correctly positioned rug)

12 a)

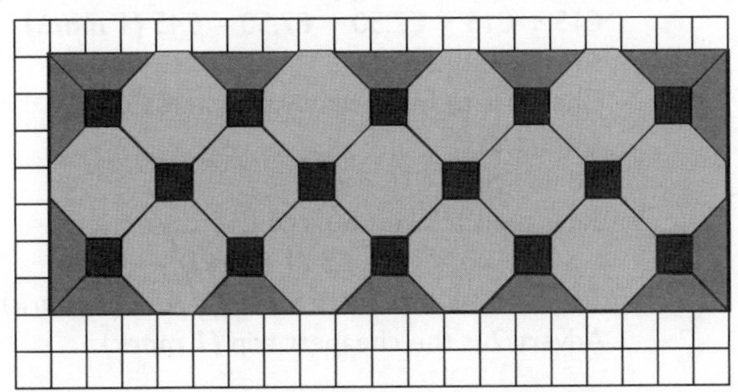

Pattern made up of stone and granite tiles with no gaps between them ***(1 mark)***. Amber tiles used to fill up any gaps at the side of the pattern ***(1 mark)***.

b) E.g. 75 ÷ 5 = 15, so each row would need 15 tiles ***(1 mark)***.
25 ÷ 5 = 5, so Vin needs 5 rows ***(1 mark)***.
5 rows of 15 tiles means he needs 15 × 5 = 75 tiles in total, so yes, he does have enough tiles ***(1 mark)***.

You could also answer this question by working out the area Vin wants to tile, and the area that the 78 tiles would cover.

c) The wall section by the sink is 0.95 m long and 0.35 m high. The wall section by the bath is 0.75 m long and 0.25 m high ***(1 mark)***. So the area of the wall section by the sink is 0.95 × 0.35 = 0.3325 m^2 and the area of the section by the bath is
0.75 × 0.25 = 0.1875 m^2 ***(1 mark)***. The total area is
0.3325 + 0.1875 = 0.52 m^2 ***(1 mark)***.
Two 1 litre tubs (0.8 m^2) cost £2.99 × 2 = £5.98.
One 1 litre tub plus one 750 ml tub (0.7 m^2) cost £2.99 + £1.99 = £4.98
Two 750 ml tubs (0.6 m^2) cost 2 × £1.99 = £3.98
(1 mark for comparing at least two combinations).
So Vin should buy two 750 ml tubs ***(1 mark)***.

Task 5 — A Trip to the Zoo (Page 92)

13 a) Mike ***(1 mark)***. (He travels 18.4 miles from Millom)

b) Closer to where Sophie lives ***(1 mark)***.

c) Length of journey = 18.4 miles
Cost of journey = 18.4 × £0.60 = £11.04 ***(1 mark for correct working, 1 mark for correct answer)***

d) Millom to Greenodd = 17.8 miles
Greenodd to Dayton/zoo = 9.4 miles
Total length of journey =
17.8 miles + 9.4 miles = 27.2 miles ***(1 mark)***
Cost of journey = 27.2 × £0.60 = £16.32 ***(1 mark)***
Difference in journey cost =
£16.32 – £11.04 = £5.28
So Mike's journey will cost him £5.28 more ***(1 mark)***.

As long as you use your answer to c) and the right method then you'll get full marks for d). It doesn't matter if you got the wrong answer in part c).

14 a) Adult ticket = £13.00
Student ticket = £9.50
£13.00 + £13.00 + £9.50 + £9.50 = £45.00 ***(1 mark)***.

b) 4 adult tickets at normal price = 4 × £13 = £52 ***(1 mark)***
4 adult tickets with voucher = £52 – 10%
10% of 52 = 10 ÷ 100 × 52 = 5.2 ***(1 mark)***
£52 – £5.20 = £46.80, which is more expensive than buying two student and two adult tickets (£45.00) ***(1 mark)***.

You need to compare the cost for 4 adults with a voucher to your answer to part a). So even if you got a) wrong you can still get full marks here.

c) £13.00 × 3 = £39.00.
3 adult admissions are more expensive than an annual pass (£35) ***(1 mark)***, so it would be worth it for him to buy one ***(1 mark)***.

15 a) Lions — 10:30
Penguins — 11:30
Giraffes — 12:00
Lemurs — 13:30
Vultures — 14:00
Monkeys — 15:00
(1 mark for including 3-5 animals at the correct times, 2 marks for including all 6 animals at the correct times. Maximum marks = 2.)

b) After the giraffe feeding and before the lemur feeding ***(1 mark)***. There will be a gap of about 1 hour 10 minutes between these sessions ***(1 mark)***.

c) No, the lemurs' and vultures' feeding sessions are now happening at the same time ***(1 mark)***.

16 a) Sandwich, drink and crisps separately:
£2.40 + £1.30 + £0.50 = £4.20
Meal deal = £3.50
Saving (difference in price) =
£4.20 – £3.50 = £0.70 / 70p ***(1 mark for correct working, 1 mark for correct answer)***.

b) Sandwich, drink and cake separately:
£2.40 + £1.30 + £0.55 = £4.25
Meal deal = £3.50
Price difference = £4.25 – £3.50 = £0.75 / 75p
So Sophie's lunch would be £0.75/75p more expensive than the meal deal. ***(1 mark for correct working, 1 mark for correct answer)***.

c) Sandwich, drink and cake separately: £4.25
Meal deal + cake = £3.50 + £0.55 = £4.05
(1 mark)
Price difference = £4.25 – £4.05 = £0.20 / 20p
(1 mark). So Jacob is right ***(1 mark)***.

Task 6 — Summer Holiday (Page 96)

17 a) i) Half-board for 7 nights in August:
£330 per adult.
Cost for Mr and Mrs Landy:
£330 + £330 = £660.
Child price = half adult price
So cost for 2 children is the same as cost for 1 adult (£330).
Total cost for family = £660 + £330 = £990.
(1 mark for correct working, 1 mark for correct answer).

ii) Budget = £1100
Cost of holiday = £990
Amount of budget left over
= 1100 – 990 = £110 ***(1 mark)***.

If you didn't get £990 for a)i) you'll have got a different answer to a)ii). You can still get the mark, as long as you use your answer to a)i) and the right method.

iii) Half-board for 14 nights in June:
£360 per adult.
Cost for Mr and Mrs Landy:
£360 + £360 = £720.
Child price = half adult price
So cost for 2 children is the same as cost for 1 adult (£360).
Total cost for family = £720 + £360 = £1080.
Yes they can afford the June holiday.
(1 mark for correct working, 1 mark for correct answer).

b) i) They need to arrive no later than 12:00 (2 hours before departure) ***(1 mark)***. So the latest train the Landys can catch is the 10:22 (which gets in at 11:17) ***(1 mark)***.

ii) They need to set off 20 minutes before the 10:22 train to give 15 minutes walking time and 5 minutes for ticket collection (5 + 15 = 20 minutes) ***(1 mark)***. So they need to set off at 10:02 ***(1 mark)***.

iii) The 11:56 arrives at the airport at 12:47 ***(1 mark)***. The Landys need to be at the airport 2 hours before 14:30 (when their flight leaves) which is 12:30 ***(1 mark)***. So catching this train will mean they are not early enough for their flight ***(1 mark)***.

c) i) Mr Landy's bag = 2.8 stones = 17.8 kg
(1 mark).
17.8 kg – 15 kg = 2.8 kg over weight limit
(1 mark).

ii) Work out everyone's bag weight in kg:
Mrs Landy's bag: 2.3 stones = 14.6 kg
Daughter's bag: 1.7 stones = 10.8 kg
Son's bag: 2.0 stones = 12.7 kg
(1 mark)
Add 2.8 kg to each bag weight:
Mrs Landy's bag: 14.6 + 2.8 = 17.4 kg
Daughter's bag: 10.8 + 2.8 = 13.6 kg
Son's bag: 12.7 + 2.8 = 15.5 kg
(1 mark)
So Mr Landy could add the excess baggage weight to his daughter's bag ***(1 mark)***.

If you got c)i) wrong then you may have got a different answer to c)ii). You can still get the marks, as long as you've used your answer to c)i) and the right method.

d) i) Buying tickets separately:
€13 + €13 + €7 + €7 = €40
Family pass = €35
Saving = €40 – €35 = €5 ***(1 mark for correct working, 1 mark for correct answer)***

ii) Buying tickets separately:
€16 + €16 + €10 + €10 = €52
Price with children's discount offer = €42
Saving = €52 – €42 = €10 ***(1 mark for correct working, 1 mark for correct answer)***

iii) Advert 1:
Adult ticket: €15
Child ticket: €7.50
Total cost for family =
€15 + €15 + €7.50 + €7.50 = €45 ***(1 mark)***
Advert 2:
Cheapest to buy a family pass at €35.
Advert 3:
Adult ticket: €16 each
Each child's ticket: 50% of €10
= 50 ÷ 100 × 10 = €5 ***(1 mark)***
Total = €16 + €16 + €5 + €5 = €42 ***(1 mark)***
Advert 2 is the cheapest trip ***(1 mark)***.

Task 7 — In the Kitchen (Page 100)

18 a) i) Butter: 1 lb = 450g ***(1 mark)***
Sugar: 1½ lb
1 lb = 450 g
½ lb = 450 ÷ 2 = 225 g.
Total = 450 + 225 = 675 g ***(1 mark)***
Flour: 2 lbs = 450 × 2 = 900 g ***(1 mark)***

ii) Milk: ½ pint = 570 ÷ 2 = 285 ml ***(1 mark)***

iii) 330 – 32 = 298 ***(1 mark)***
298 × 5 = 1490 ***(1 mark)***
1490 ÷ 9 = 165.55
So 330 °F = 165.55 °C (accept answers between 165 and 166 °C) ***(1 mark)***

b) i) The amounts in the recipe are for 36 cakes, Katie wants to make 12 cakes.
$36 \div 12 = 3$
So the amounts all need to be divided by 3.
Butter = $375 \div 3 = 125$ g
Sugar = $750 \div 3 = 250$ g
Flour = $600 \div 3 = 200$ g
Eggs = $6 \div 3 = 2$ eggs
Milk = $360 \div 3 = 120$ ml
(1 mark for attempting to divide each amount by 3, 1 mark for 3-4 correct answers, 2 marks for all 5 answers correct. Maximum marks = 3.)

ii) $48 = 12 \times 4$.
Sugar for 12 cakes = 250 g.
So for 48 cakes = $250 \times 4 = 1000$ g (1 kg).
(1 mark)
Flour for 12 cakes = 200 g.
So for 48 cakes = $200 \times 4 = 800$ g.
(1 mark)

There are different ways to work out the answers to this question. As long as your answers are correct, you'll still get the marks.

c) i) No (she has added $3\frac{1}{2}$ oz).
She needs to add 1 oz more ***(1 mark)***.

ii) $6\frac{1}{2} + 4\frac{1}{2} + 4 = 15$ oz ***(1 mark)***

You can also work out this answer by counting the number of ounces of each ingredient off on the scale.

iii) Butter : flour = 1 : 4
So the amount of flour is 4 times the amount of butter.
$3\frac{1}{2}$ oz $\times 4 = 14$ oz of flour.
(1 mark for correct working, 1 mark for correct answer)

d) i) 100 g + 50 g + 20 g ***(1 mark)***

ii) The diagram shows that 4 carrots weigh 300 g.
You need 300 g more carrots to make 600 g in total.
That is an extra 4 carrots.
$4 + 4 = 8$ carrots in total.
(1 mark for reading diagram, 1 mark for correct working, 1 mark for correct answer)

iii) There's 125 ml of milk in the jug. Maria needs 150 ml. $150 - 125 = 25$ ml.
So Maria needs to add 25 ml of milk to the jug.
(1 mark for correct working, 1 mark for correct answer)

Task 8 — Tony's Bookshop (Page 104)

19 a)

Day	Tally
Monday	𝍸
Tuesday	𝍸 III
Wednesday	IIII
Thursday	𝍸 IIII
Friday	𝍸 𝍸 I
Saturday	𝍸 𝍸 𝍸 I

(1 mark for correct column headings, 1 mark for tallying 4-5 days correctly, 2 marks if all 6 days tallied correctly. Maximum marks = 3.)

b) i) Mean:
$10 + 11 + 8 + 13 + 16 + 20 = 78$
$78 \div 6 = 13$ ***(1 mark for correct working, 1 mark for correct answer).***

ii) Range: 8, 10, 11, 13, 16, 20 = $20 - 8 = 12$
(1 mark for correct working, 1 mark for correct answer).

iii) For example:

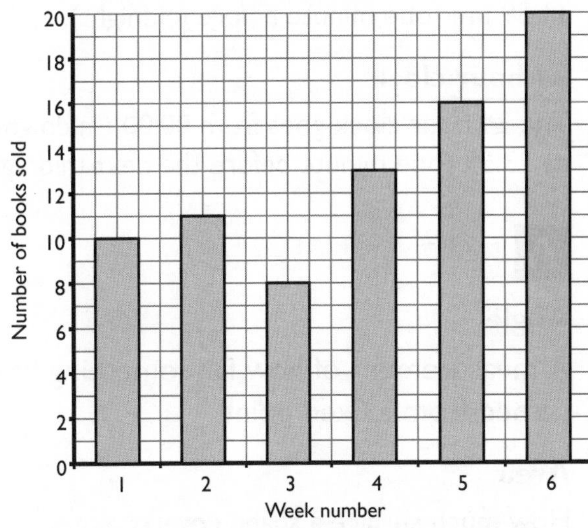

(1 mark for choosing an appropriate chart or graph, 1 mark for choosing a sensible scale, 1 mark for correct labelling of axes and a title, 1 mark for correctly plotting data).

This is one possible chart you could draw from the information provided. If you've drawn something a bit different, don't worry. As long as what you've got is correct, you'll still get the marks.

c) i) June books = $7 + 8 + 3 + 7 + 4 + 4 = 33$
July books = $5 + 9 + 7 + 2 + 5 + 2 = 30$
Tony received the most new books in June.
(1 mark for correct working, 1 mark for correct answer given)

ii) Action: $7 + 5 = 12$
Crime: $8 + 9 = 17$
Romance: $3 + 7 = 10$
Historical: $7 + 2 = 9$
Sci-Fi: $4 + 5 = 9$
Horror: $4 + 2 = 6$
So Tony received the most books from the crime category.
(1 mark for correct working, 1 mark for correct answer)

iii) June: 7 action + 8 crime = 15
July: 5 action + 9 crime = 14
So the total number of action and crime books received was highest in June.
(1 mark for correct working, 1 mark for correct answer)

Glossary

12-hour clock

The 12 hour clock goes from 12:00 am (midnight) to 11:59 am (one minute before noon), and then from 12:00 pm (noon) till 11:59 pm (one minute before midnight).

24-hour clock

The 24 hour clock goes from 00:00 (midnight) to 23:59 (one minute before the next midnight).

Angle

A measurement of how far something has turned from a fixed point.

Area

How much surface a shape covers.

Average

A number that summarises a lot of data.

Axis

A line along the bottom and up the left-hand side of most graphs and charts.

Bar Chart

A chart which shows information using bars of different heights.

Capacity

How much something will hold. For example, a beaker with a capacity of 200 ml can hold 200 ml of liquid.

Certain

When something will definitely happen.

Decimal Number

A number with a decimal point (.) in it. For example, 0.75.

Even Chance

When something is as likely to happen as it is not to happen.

Formula

A rule for working out an amount.

Fraction

A way of showing parts of a whole.
For example: ¼ (one quarter).

Frequency Table

A tally chart with an extra column that shows the total of each tally (the frequencies).

Function Machine

A way of showing formulas that have more than one step.

Impossible

When there's no chance at all of something happening.

Length

How long something is. Length can be measured in different units, for example, millimetres (mm), centimetres (cm), or metres (m).

Likely

When something isn't certain, but there's a high chance it will happen.

Line Graph

A graph which shows data using a line.

Line of Symmetry

A shape with a line of symmetry has two halves that are mirror images of each other. If the shape is folded along this line, the two sides will fold exactly together.

Mean

A type of average. To calculate the mean you add up all the numbers and divide the total by how many numbers there are.

Mileage Chart

A type of table that shows you the distance between different places.

Negative number

A number less than zero. For example, -2.

Percentage

A way of showing how many parts you have out of 100. So twenty percent (20%) is the same as 20 parts out of 100.

Perimeter

The distance around the outside of a shape.

Pictogram

A chart that uses pictures or symbols to show how many of something there are.

Pie Chart

A circular chart that is divided into sections (that look like slices of a pie). The size of each section depends on how much or how many of something it represents.

Plan

A diagram to show the layout of an area. For example, the layout of objects in a room.

Probability

The likelihood (or chance) of an event happening or not.

Protractor

A piece of equipment used to measure angles in degrees (°) up to 180°.

Range

The difference between the biggest and smallest numbers in a data set.

Ratio

A way of showing how many things of one type there are compared to another.
For example, if there are 3 red towels to every 1 white towel then the ratio of red to white towels is 3 : 1.

Symmetry

See line of symmetry.

Table

A way of showing data. In a table, data is arranged into columns and rows.

Tally Chart

A chart used for putting data into different categories. You use tally marks (lines) to record each piece of data in the chart.

Tessellation

When shapes are put together in a pattern to leave no gaps.

Unit

A way of showing what type of number you've got. For example, metres (m) or grams (g).

Unlikely

When something isn't impossible, but it probably won't happen.

Volume

The amount of space something takes up.

Weight

How heavy something is. Grams (g) and kilograms (kg) are common units for weight.

Index